# A New Paradigm for Greek Agriculture

Kostas Karantininis

# A New Paradigm for Greek Agriculture

Kostas Karantininis
Swedish University of Agricultural Sciences
Uppsala
Sweden

ISBN 978-3-319-59074-5     ISBN 978-3-319-59075-2   (eBook)
DOI 10.1007/978-3-319-59075-2

Library of Congress Control Number: 2017943495

Cover illustration: © saulgranda/Getty

Printed on acid-free paper

This Palgrave Macmillan imprint is published by Springer Nature
The registered company is Springer International Publishing AG
The registered company address is: Gewerbestrasse 11, 6330 Cham, Switzerland

Most of the research behind this publication has been funded by the Task Force Greece (TFGR) under the European Commission (EC) Service contract VC/2013/1171 "to provide technical assistance to the Greek authorities on rural development and fishery sector with an assessment of new opportunities for agricultural sector and the rural areas and the necessary reforms for full exploitation of its potential". The results were first published in the report ASSISTANCE TO NEW OPPORTUNITIES FOR AGRICULTURE IN GREECE (ANOAG), and in Greek as ΝΕΕΣ ΕΥΚΑΙΡΙΕΣ ΓΙΑ ΤΗ ΓΕΩΡΓΙΑ ΣΤΗΝ ΕΛΛΑΔΑ, both in 2014. The present publication is an independent report and reflects only the opinion of the author and not those of the SLU or of the European Commission nor those cited, or interviewed anonymously carry any responsibility for the information, the opinions and recommendations presented in this publication.

# Acknowledgements

This book is the amalgam of research and discussions with numerous knowledgeable and insightful people, who generously provided their time, energy, knowledge and passion for their work and Greek agriculture. The references provided at the end of each chapter are only a sample of the sources that have been sought after for this research. I also provide here some names that still remain in my memory. Many more are omitted and is only because they faded my memory and are lost in my notes. I ask them to forgive me, although, as they will find after reading this book, many of their ideas and information are here. The usual caveat holds: the author of this book is the sole responsible for any errors or omissions.

First and foremost I wish to thank the Task Force Greece (TFG), for funding and entrusting me with this research. The TFG with the Embassy of The Netherlands in Athens provided assistance and allowed me the freedom to express all my ideas and make my own mistakes. Director of TFG Georgette Lalis and the invaluable Vassilis Kappas were most helpful. Vassilis was a model employee for the TFG, was my guardian angel and assistant during the research process. Tassos Haniotis of DG-Agri in Brussels, was always there for me, insightful, critical and resourceful. I was assisted by many knowledgeable and extremely helpful people at the Greek Ministry of Rural Development and Food (GMRDF), in various Departments, too many to mention—thank you all. The staff at Piraeus Bank was very helpful, insightful and transparent. In Crete, the late Mr. Alkinoos Nikolaidis, and his successor, director of MAICH, George Baourakis provided insights and contacts. George Demetriadis of BIOLEA, and the invaluable force of INKA, "the president" Panagiotis Alexakis were most resourceful. Wine makers, Stelios Boutaris, Evangelos Gerovassiliou, Aggelos Rouvalis, provided fuel for thought, ideas and encouragement. Chef Elias Mamalakis was extremely generous, and provided with a wealth of insights in agri-food. I cannot thank enough Kostas Kastrinakis, Giouli Doxanaki, experts in Greek agri-food. In Larissa, I had the privilege to meet and receive insights from two leaders of modern cooperatives in Greece, Thanassis Vakalis of THESGALA and Panagiotis Kalfountzos of THESGH, as well as Adonis Malamis, their staff and farmer members of their pioneering cooperatives. Agronomists Giannis Tsoumanis, Vassilis Chalkidis, Thanassis Tamparopoulos and Veterinarian and banker Christos Makris, thank you. Several leaders from Greek agri-food contributed with critical insights and data, Nikos Efthymiadis of Redestos, Efthymis Tsimpidis of Green Bay, several executives of DELTA, and many others, provided insights and critical thoughts.

I met many farmers in this process and of course during my career as researcher and my previous life as farmer. Especially my compatriots and friends from my village Trikala Imathias have helped and taught me a lot about life and agriculture. My cousin Christos Ouvaloudis, farmer and jack-of-all-trades, kept me grounded when I was speaking "professorial". Special mention to George Pegioudis, agronomist, farmer and beacon of organics in Greece. I used George to bounce ideas and check relevance from somebody "from the ground". George was always there, critical, and to the point. Thank you George for your generosity, your trust and friendship and for the countless hours of Skype talks.

Many colleagues who over the last years helped me organise and participated in the yearly conferences at Trikala Imathias have contributed, often without knowing, to the development of this manuscript. The usual suspects: Kostas Katrakylidis, Nikos Varsakelis, Ioannis

Kyritsis, as well as Kostas Mattas, Christos Kamenidis, of the Aristotle University of Thessaloniki, Eftychios Sartzetakis of the University of Macedonia, Aspasia Vlachvei of TEI of Western Macedonia, Rania Notta of TEI Thessalonikis, Dimitris and Machi Tsoukala from Anavra, Stavros and Theo Benos of DIAZWMA, Vangelis Vergos of the American Farm School, Vangelis Divaris of FILAIOS, Kostas Polymeros, Alexis Koutsouris and several other colleagues from ETAGRO and many other speakers and participants. My friends, Kalmpoukias, Potkas, Mokas, Stelios, George Mourtz, Tsiopis, Tom Fisher: you know. The mayor of Trikala, Maria Papaioannidou, Simos Tamoglou who provided IT coverage, and the poster team, Giota Staktari and Liana Zachariadou, thank you.

While working on the manuscript, many times I have seen the sun rise behind the cherry tree in our garden in Copenhagen. My wife and life companion, Sevie Chatzopoulou, was always there, patient and encouraging, insightful and critical—an excellent discussant. I owe a lot to Sevie for this and for most of my work. Many ideas in this book are Sevie's as much as they are mine, only she would have expressed them with better precision, and she would have avoided most of the errors I made. I need also to thank and apologise to my two children, Marilena and Adonis for their love and for the many hours I took away from them, writing and travelling.

I have learned a lot about life and agriculture while I was farming in Trikala Imathia, Greece. My parents, Marika and Adonis Karantininis, taught me all I know about growing peaches and raising pigs, and especially how to put them to market and make an honest living out of a small family agribusiness. I still learn from my father, but unfortunately my mother cannot teach me and love me no more. I lost her while preparing this manuscript.

It is to the memory of my mother, Marika Karantinini that I dedicate this book.

# Contents

# Abbreviations

| | |
|---|---|
| ADI | Agricultural Development Institute |
| AES | Agriculture Extension Services |
| AFFI | Agri Food Firm Incubator |
| AFS | American Farm School |
| AGEAC | Agriculture Extension and Advisory Centres |
| AGFBC | Agriculture Food Business Centre |
| AGOMBUD | Agriculture Ombudsman |
| AGROCERT | (О.П.Е.Г.Е.П.) National organization of Agricultural Products Certification |
| AQIPO | Aquaculture Inter-Professional Organization |
| AQRDF | Aquaculture Research and Development Fund |
| AVS | Agricultural Vocational Schools |
| CDFM | Centre for Development of Farmers' Markets |
| CIHEAM | International Centre for Advanced Mediterranean Agronomic Studies |
| COAFRED | Committee on Agriculture Food Research and Education |
| COMAGRI | European Parliament Committee on Agriculture and Rural Development |
| EFET | (Е.Φ.Е.Т.) Hellenic Food Authority |
| ESMA | Extension Service of the Ministry of Agriculture |
| FIP | Food Innovation Park |

| | |
|---|---|
| GAEUOF | Greek Agriculture EU Office |
| GFLH | Greek Food Logistics Hub |
| GIPAFC | Greek Inter-Professional Agriculture and Food Council |
| GMRDF | Greek Ministry of Rural Development and Food |
| GOMM | (ΕΛ.Ο.ΓΑ.Κ.) Greek Organization of Milk and Meat |
| GREFEX | Greek Food Experience |
| HAERC | Hellenic Agriculture Extension Research Centre |
| HAFRDF | Hellenic Agricultural and Food Research & Development Fund |
| HFBF | Hellenic Food Brand Franchise |
| HIAFR | Hellenic Institute of Agriculture and Food Research |
| IOF | Investor Owned Firm |
| IPO | Inter-Professional Organizations |
| KEGE | Local Training Centres of DIMITRA |
| LFMC | Local Farmers' Market Cooperative |
| MAICH | Mediterranean Agronomic Institute of Chania |
| NAGREF | (ΕΘ.Ι.ΑΓ.Ε.) National Agriculture Research Foundation |
| NFFMC | National Federation of Farmers' Markets' Cooperatives |
| NORC | National Olive Research Centre |
| OAVETE | (ΟΓΕΕΚΑ) Organization of Agricultural Vocational Education Training and Employment |
| OLIPO | Olive Inter-Professional Organization |
| OLRDF | Olive Research and Development Fund |
| OPEKEPE | Organization of Payments and Control of Payments and Guarantees |
| PASEGES | Panhellenic Confederation of Agricultural Cooperatives |
| PPCFARD | Permanent Parliamentary Committee of Food Agriculture and Rural Development |
| PSFARD | Permanent Secretary of Food, Agriculture and Rural Development |

# List of Figures

# List of Tables

# 1

# Introduction

This book argues for a new paradigm for the Greek agri-food industry. The book employs a value chain approach to the problem and proposes a paradigm that incorporates the entire agri-food industry; it relates agri-food to tourism, logistics, research, education and integrates organizational, political and institutional arguments. This book is not a theoretical contribution, nor an exhaustive empirical analysis of the Greek agri-food sector. The *New Paradigm for Greek Agriculture* aims to become a platform for discussion around the agri-food sector of Greece with the hope to raise similar discussions in other countries with similar to Greece's diverse and rich agri-food sectors. The book has five main points of departure:

The first point of departure for this book is the resilience of Greek agriculture and the agri-food sector. During the last four decades, agriculture had a declining contribution to the country's GDP, from more than 30% in the 1970s to less than 5% before the 2008 financial crisis. The contribution of agriculture increased to 5.6% in 2015.

The second point of departure is the unrealized potential of the food industry. Food processing accounts for about 25% of the value added of the total manufacturing sector and employs 22% of the total manufacturing

© The Author(s) 2017
K. Karantininis, *A New Paradigm for Greek Agriculture*,
DOI 10.1007/978-3-319-59075-2_1

labour force. However, the value added by the Greek agri-food firms lags behind its own true potential and its competitors in Europe and elsewhere, leaving great room for improvement.

A third point of departure is the advantage of other sectors, mainly tourism and logistics, and their potential link to agri-food. Greece is visited by more than 20 million tourists annually—twice the country's permanent inhabitants. Besides that this adds to the food demand, these visitors can be targeted for the promotion of Greek food products so they continue the consumption after their return to their home countries. Greece is also turning into a main logistics hub for Europe, which provides a significant potential for Greek food products. It poses also a potential challenge by exposing Greece to competition with cheaper food imports.

A fourth point of departure is the diversity of Greek agriculture which does not favour large-scale industrial production, while it offers potential for broad differentiation and variety. The variety is due to a mountainous and island geography, a multitude of microclimate zones and a long historical and cultural path dependency.

The fifth point of departure is that the main constraints for the growth of Greek agriculture are institutional in nature. Both the private sector and the state institutions are poorly organized and not coordinating. There is a lack of a well-organized farmer's cooperative sector, while the political governance of the ministry and related administrative bodies, such as training, extension and research, are disconnected.

The book is organized in six chapters. After this introductory chapter, the second chapter provides a background and overview of Greek agriculture, showing the structural composition of agricultural production. Both size and geographical structural distribution exhibit a fragmented sector, dominated by small farms, mostly in western and southern Greece.

Chapter three is on policy and governance. It starts with an introduction to the Common Agricultural Policy (CAP), its role in Greek agriculture and its implementation. The main emphasis is on the structure of payments which under the new direct payments regime follow closely the distribution of farms. The discussion on governance and the political structure of the Ministry of Rural Development and Food (GMRDF) shows a fragmented and unstable governance, where ministers of agriculture change too often, while the governance is centralized around the ministry and is at the same time disconnected from the sector. The extension and advisory service suffers greatly, especially after the 1981 EU accession, which converted the majority of advisors and agronomists into "pencil pushers". The vacuum was filled by private firms, mainly those affiliated to large agrochemical corporations.

Chapter four discusses markets for inputs, land, credit, agrochemicals and output markets, focussing on the role of retail. It is not surprising that CAP payments are capitalized on land prices, while Greece is surrounded by Balkan countries with cheaper agricultural land. Hence, on a pure cost basis, Greek farmers will find it difficult to compete with their Balkan neighbours on extensive agriculture. The credit market with relatively high interest rates, mainly due to the structure of the banking sector and the ongoing financial crisis, is a burden on cash flow and overall growth for agriculture. The retail industry is highly oligopolistic, as in most EU countries, squeezing further the margins to farmers.

Chapter five examines in more detail, three subsectors, fruits and vegetables, aquaculture and olive oil. They are used to illustrate further the common problems of fragmented farms, high credit costs, lack of coordination in the agri-food chain, and poor governance.

The necessary institutional innovations for the new paradigm are presented in chapter six. The framework is based on the distribution of economies of size along the chain, the importance of the experience economy and the scaling up. The fundamental structure is the Greek Inter-Professional Agri-food Council (GIPAFC)—a square pyramid with inter-branch organizations, cooperatives, research and other sectors on its four pillars. The governance structures include the entire spectrum from markets to hierarchies, including collective action and integrated farmers' markets. The GIPAFC is linked to policy governance organizations and to the entire agri-food upstream and downstream.

The first draft of the new paradigm for Greek agriculture as presented here, is perhaps incomplete, but includes an economic rationale, the institutional background, the organizational structure and its political governance - it is thus "new". The new paradigm will require a long process of consultations and adaptation among its stakeholders: farmers, farmer organizations, industry, government and other sectors. What is needed is a target and a guideline.

# 2

# Overview of Greek Agriculture

**Abstract** Greek agriculture has shown a remarkable resilience during the financial crisis of 2008–today. Agriculture output increases, while all other sectors' output fell by a gross 30%. Greek agriculture is dominated by small farms scattered over a diverse territory. The structure of Greek agriculture reflects the unique geography and climatic conditions, the historical path and the cultural diversity of the country. The challenge is to turn this diversity into a competitive advantage via reorganization following a new paradigm.

**Keywords** Greek agriculture · Farm structures · Agricultural crisis

## 2.1 Agriculture Anti-cyclical

After Wall Street imploded in 2008, Greece dived into a deep recession. The Greek government announced in October 2009 that it had been understating its deficit figures for years. This raised alarm about the soundness of Greek finances and Greece was shut out from the

© The Author(s) 2017
K. Karantininis, *A New Paradigm for Greek Agriculture*,
DOI 10.1007/978-3-319-59075-2_2

international financial markets. By the spring of 2010, the country was on the verge of bankruptcy, threatening to pull the entire world financial system into a new financial crisis. In a collaborative action to avoid the worse, three institutions—the so-called troika—the International Monetary Fund (IMF), the European Central Bank (ECB) and the European Commission (EC), decided to bail out Greece with collective international funding that reached eventually more than €240 billion.

The lenders asked for austerity measures, deep budget cuts, tax increases and a general restructuring of public administration, by streamlining the government, fighting tax evasion and easing business processes. The terms of the agreement were set out into three consecutive memoranda of understanding, signed between the Greek government and the "troika"— as the three institutions are known in the Greek crisis jargon.

Greece was certainly not the only country that was hit by the crisis. Cyprus, Portugal and Ireland also required an intervention by the international institutions; however, one after the other, they managed to exit the crisis, while Greece still remains under surveillance by the four so-called "institutions" (The European Stability Mechanism joined the troika in 2015) and functioning within three consecutive memoranda.

Greece is in a crisis path for almost a decade now, undergoing a devastating recession, which has caused the country a loss of more than one-third of it GDP. This has hit almost every single citizen and every single sector of the Greek economy. Yet there is one sector that has exhibited a remarkable resilience: agriculture.

Greek agricultural sector ranks tenth among EU-28 member countries in terms of the total value of agricultural production at EURO 9.6 billion in 2015 (EUROSTAT). The total value of agricultural production (not including subsidies) follows an upward trend, which continues even during the years of the financial crisis (2008 and onwards) (Fig. 2.1). However, the Agricultural Goods Output (AGO) of agriculture (production less intermediate consumption of variable inputs) has been decreasing since 2005 (after the 2003 CAP reform), although at a slower rate during the crisis with indications of upward movement after 2013 (Fig. 2.1). The decline in agricultural GDP can be explained partly by the increase in the costs of production.

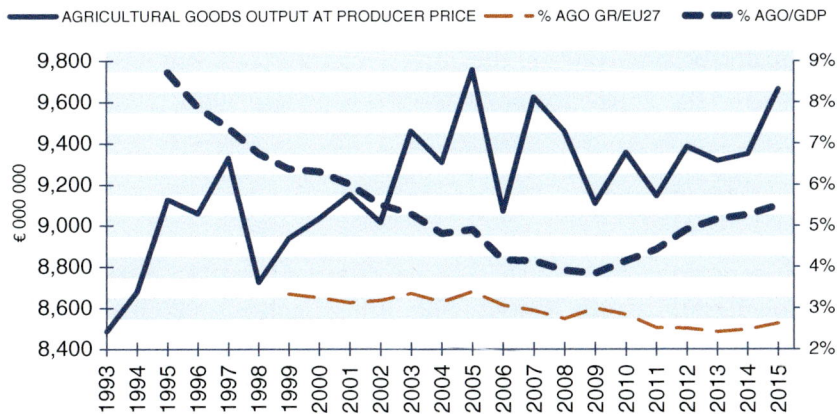

**Fig. 2.1** Agricultural production as % of Greek GDP and % EU-27. *Source* EUROSTAT, own calculations

The importance of Greek agricultural production, both relative to the rest of the EU and in terms of contribution to Greek GDP, is overall shrinking, although the picture is changing during the crisis. Greek agriculture has shown a poor performance relative to the rest of EU. The share of Greece's agricultural output in EU-27 decreased from 3.3% in 1993 to 2.6% in 2015. Similarly, the growth of farm income, both in terms of income per worker (indicator A) and in terms of entrepreneurial income (indicator B), has followed disparate paths from similar EU indicators, especially after 2009 (Eurostat).

The share of agriculture in Greek GDP decreased from 8.7% in 1995 to 5.5% in 2015. Notice, however, the relative anti-cyclical behaviour of the agricultural sector: during the years of the Greek financial crisis (since 2008), the contribution of Greece's agricultural production to the country's GDP has increased, mainly due to a more than 30% collapse of GDP. The share of agriculture increased from 3.8% in 2008 to 5.5% in 2015. This is true in comparison with the rest of the Greek economy— not true however relative to the rest of the EU.

The fact that a sector remains strong in the middle of chaos where the country has lost one-third of its national income is a very positive trait of the Greek agricultural sector as a whole. This resilience needs to be understood, improved and scaled up. Agriculture and food could potentially become one of the drivers of future growth in the Greek economy. This, however, may require a paradigm shift.

## 2.2   Structures

Greek agriculture is dominated by small farms, and the distribution of land ownership is very skewed. The utilized agricultural area (UAA) in Greece was 2.8 m. ha in 2013 (excluding grassland).[1] This is about 21% of the total country area (40% if grassland is included). Whereas land in Greece is distributed relatively evenly between small, medium and large holdings (Fig. 2.2), ownership is very much concentrated. While 10.5% of the UAA is distributed among very small farms (with less than 2 ha), these small holdings comprise more than half of the total number of farms (53.6% or 347,950 farms out of the total 648,610 in 2013). Most farms (approximately 80%) are below the country average (4.6 ha in 2013), while about 95% of the farms would fall below the EU-27 average (14.6 ha).

A small number of farm operators (470 holdings, or 0.07%) operate farms larger than 100 ha, which constitute 3% of total UAA (Fig. 2.2). A large reduction in total UAA occurred between 2007 and 2010—a decrease of 529,350 ha, −16.25%, whereas this is stabilized in 2013 at 2,754,300 ha. Similarly, the total number of holdings is decreased by 115,560 farms, or −14.9%, between 2007 and 2010 and by 11,440 farms between 2010 and 2013. The largest absolute decrease is in the smallest farms (a decrease of 87,170 farms, or −14.1%). On the other hand, the number of the largest farms (>100 ha) increased by 40 (or +14.3%) between 2007 and 2010 and by 150 holdings (+46.8%) between 2010 and 2013. The area cultivated by the largest farms increased by 12.6% and 22.8% between 2007 and 2010 and 2010 and 2013, respectively. Entry in this category was by farms at the bottom of the scale (100 ha), as indicated by the decrease in the average size of the >100 category from 215.75 ha to 212.56 ha and 177.8 ha, for 2007, 2010 and 2013, respectively. The dynamics of the farms size distribution show a decrease in holdings and UAA by holdings of less than 10 ha.

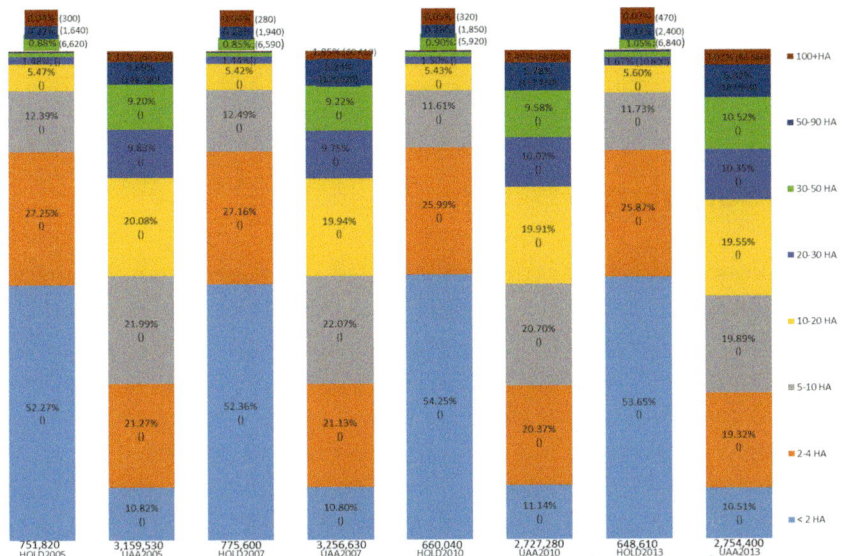

**Fig. 2.2** Distribution of utilized agricultural land (UAA) and number of holdings (%). *Source* EUROSTAT, own calculations

The overall average size of farm did not change much in this period but fluctuated: 4.2 (2005), 4.19 (2007), 4.13 (2010), 4.25 (2013). Concerning these figures, one has to be careful since data definitions may vary and also be aware that it is common practice, usually for tax and other reasons, with farmers in Greece to register holdings under other family members. Therefore, the number of holdings and the average farm size are not representative of the actual situation. The average farm sizes are actually larger, although we cannot tell with certainty by how much.

Due to geography and certain historical path dependence, the distribution of land varies between areas in Greece. In 2013, there were 347,950 (53.7%) small farms in Greece with size less than 2 ha (Fig. 2.2). The distribution is very unequal and skewed between regions. In the entire country, 10.5% of the land is held by farms with less than 2 ha (excluding permanent grasslands). In some regions, more land is held by small farmers; this is true for mountainous areas and islands. In the Ionian Islands, for example, 37.4% of the land is small farms, and 29.1% in Attiki (EUROSTAT).

Most of the small farms (<2 ha), a cumulative 52.9% of all small farms in Greece, are in four regions: DytikiEllada (51,430 farms), StereaEllada (36,420 farms), Peloponnisos (48,040 farms) and Kriti (53,620 farms).

## 2.3  Summary

The number, size and distribution of small farms are a complex matter. We only touch this issue briefly in this book due to time and size limitations. We look only at the regional dimension of the structure of farms; however, it is important to analyse structures for various farms activities and crops, which goes beyond the scope of this book. However, any serious agricultural and rural development policy must look at this issue very carefully and more resources need to be dedicated to further analysis and understanding of small farms, and their role in development and restructuring. Small farms differ from one area to the next, and between crop, or livestock activities. A single policy that fits all small farms is neither desirable nor possible. Instead, policy-makers must design customized and specialized policies that deal with the complexity and special needs of each region and farm activity.

Often, small farms are considered an impediment to development and growth. Policy-makers, however, should pay attention to the fact that while small farms may often - but not always - exhibit diseconomies of scale, they offer variety and product differentiation. Furthermore, in some mountainous regions, small farms are the only feasible structure. One important factor that is stressed very much in this report is that the largest gains in economies of scale are not necessarily at the primary production level, but mostly downstream and upstream the farm. The logistics and distribution system processing, as well as farm inputs and research, are operations that exhibit large economies of scale. If the value chain is integrated and coordinated appropriately, and the benefits of scale are distributed evenly along the chain, then small farms will not only simply survive, but they could contribute greatly to variety and product differentiation. In this way, small farms can add income to families and growth and development in the rural areas. Without an integrated value chain, however, the small farms are doomed to become extinct, with very severe negative consequences to rural development and the continuation of social, economic and cultural life in the rural areas. This does not mean that a policy for Greek agriculture and food should aim at maintaining every single farm, irrespective of size and level of efficiency and capabilities of its operator. Instead, the agricultural policy should aim at developing a structure in the value chain and should provide support that offers opportunities to all farms, small or large, to grow and add value to the system and income to its owner and to the local economy. Farms, small or large, should not be supported in perpetuity, if their operators are not able to, or are not willing to add value and adjust to the needs of the demand and to respond to the opportunities and incentives offered by a well-structured and sustainable value chain.

Careful planning with strategic targeting of funds, for the development of a sustainable value chain, can lead to the sustainability of small farms. Any policy, however, should aim at providing balanced opportunities to all involved.

## Note

1. Structures exclude "Permanent grassland and meadow". In 2010, Greece had to report all its "common land area" which were added to structures mostly as grasslands under a single ownership (owned by "the local community"). "Permanent grass land and meadow" was 546, 440 ha (76,350 holdings) in 2005, 819,610 ha (78,520 holdings) in 2007, whereas it increased to 2,450,240 ha (56,830 holdings) in 2010 and 2,102,380 ha (54,990 holdings) in 2013. These figures are excluded from structures reported here. (Statistics Explained (http://epp.eurostat.ec.europa.eu/statistics_explained/)—22/05/2014- 15:49)

# 3

# Policy Governance

**Abstract** Greek agriculture is dependent on cap subsidies, which have shaped production to a certain extent. The structure of CAP payments, especially under the new CAP, reflects the structure of farms. Greece has been charged often and with large amounts for mishandling of CAP payments. The political governance of the agricultural sector is unprofessional, inefficient and highly centralized. Extension and vocational training is also highly inefficient and underfunded. Agricultural research in Greece is above EU average, but is not linked to the sector.

**Keywords** Greece and CAP · Political governance · Structure of CAP payments · Greek agricultural extension

## 3.1 CAP: Common Agriculture Policy

The accession of Greece in the EU in 1981 and the consequent implementation of the Common Agricultural Policy (CAP) have had significant ramifications on growth and structure of the country's

© The Author(s) 2017
K. Karantininis, *A New Paradigm for Greek Agriculture*,
DOI 10.1007/978-3-319-59075-2_3

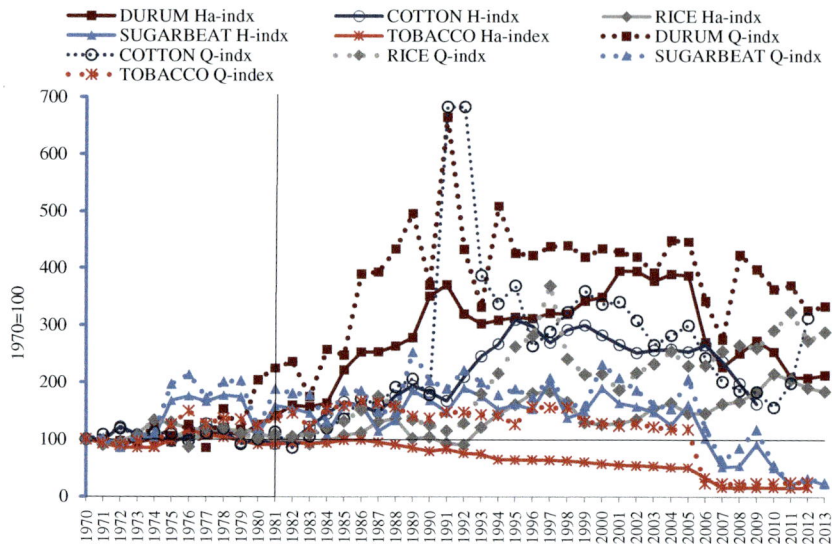

**Fig. 3.1** Index of Greek production and utilized area of selected crops (1970 = 100). *Source* EUROSTAT, own calculations

agricultural sector. Selecting only a few crops, it is striking—and not at all surprising—that the production of crops heavily subsidized under CAP has increased significantly. The utilized area under durum wheat, and cotton, for example, has more than tripled within the first decade after the accession (Fig. 3.1). On the contrary, tobacco and sugar beets are almost being extinct.

### 3.1.1   Structure of CAP Payments

A large portion of agricultural receipts come from CAP subsidies (Fig. 3.2). About 25% of agricultural product receipts are direct payments and export refunds—a total of more than €2 billion annually (€2,315,248.9 in 2012). When compared to the net value added of agriculture (NVA)—i.e. when variable costs of production are subtracted —the contribution of subsidies reaches 45–50% (Fig. 3.2). The overall dependence of EU-28 agricultural revenue on CAP funds is about

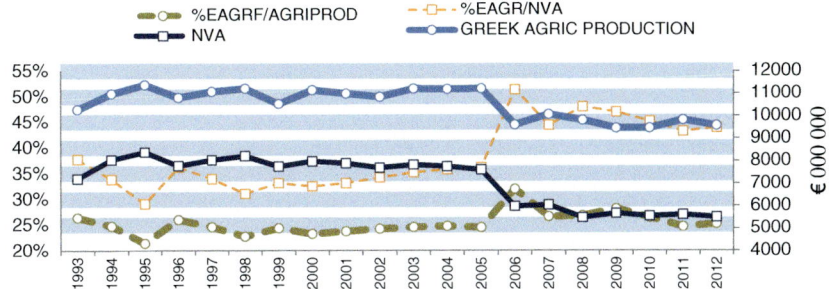

**Fig. 3.2**  Value of agricultural products and EAGF payments—Greece 1993–2012.
*Source* EUROSTAT, EAGF, own calculations—Excludes Rural Development

12%, whereas 28% of the EU-28 GVA depends on CAP payments
(EAGF financial reports and EUROSTAT).

According to the current structure of CAP financing for Greece,
78.9% (€2.3 billion in 2012) are direct payments and 21.7% (€670
million) are finances for rural development (Fig. 3.3). The plan for the
next 6 years (2015–2020) is for a gradual reduction in total financing to
Greece to €2.544 billion by 2020. A larger portion (23.4%) will be rural
development financing. In 2015, total CAP financing for Greece is
expected to decrease by 6% (based on the ceiling for 2014), and by
approximately 1% annually for the years until 2020. The total funds
expected to flow into Greece for the 6 years (2015–2020) are a total of
€15.6 billion. The bulk of these payments, €12 billion (67%), will be

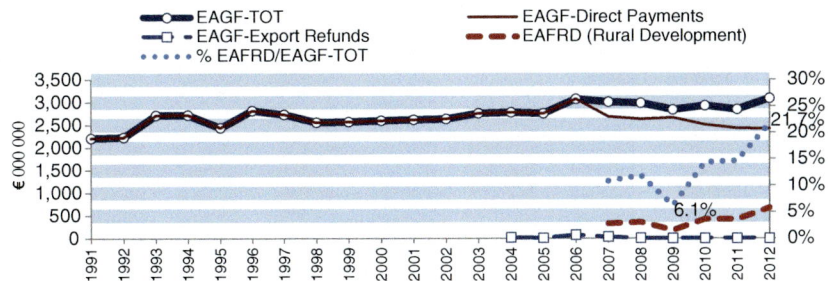

**Fig. 3.3**  Direct payments (EAGF) and rural development finances (EAFRD)—
Greece. *Source* EUROSTAT, EAGF, EAGRD, own calculations

direct payments and €3.6 billion (23%) will be funds for rural development (EAGF financial reports and EUROSTAT).

## 3.1.2  Distribution of CAP Payments

The distribution of CAP payments is not equitable. Those beneficiaries who receive more than €20,000 direct payments annually are 1.75% (12,714 beneficiaries) and receive 17.21% of the total transfers (Fig. 3.4).

We have also calculated the distribution of CAP payments as Lorenz curves and Gini coefficients (Karantininis 2014). In 2012, the Gini coefficient of direct CAP payments was 0.61, a slight improvement from 0.63 in 2005 and 0.68 in 2007. The distribution of CAP payments is very similar to the overall distribution of land. The Gini for land (excluding grasslands) was 0.62 in 2010; similarly, the Gini of CAP payments was 0.61 in 2011 and 2012. It is our understanding that the

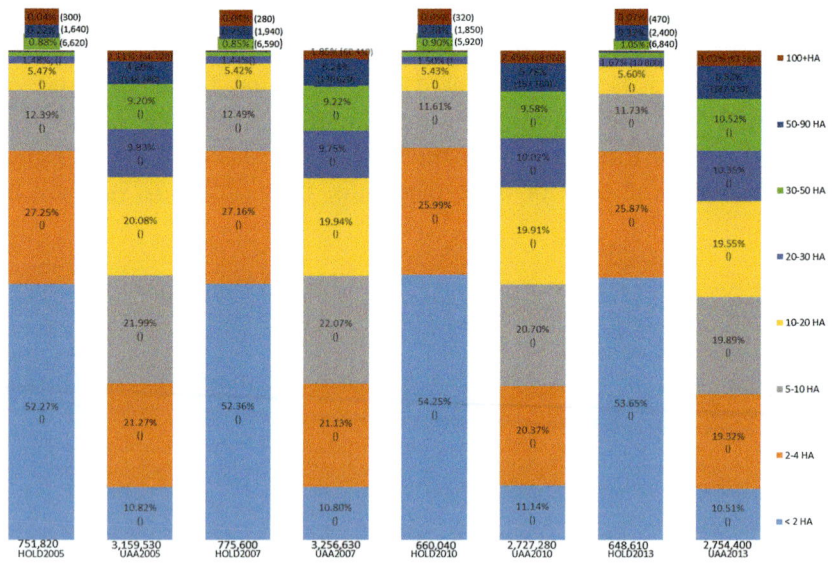

**Fig. 3.4**  Distribution of utilized agricultural land (UAA) and number of holdings (%). *Source* EUROSTAT, own calculations

distribution of payments is even more skewed, since there are many beneficiaries who split the benefits between family members. A more accurate accounting would show that there is room for improvement in the distribution of payments from top beneficiaries in Pillar I towards Pillar II, as Regulation EU 1305/2013 allows.

### 3.1.3 Implementation of the CAP

The implementation of the CAP suffers both on the direct payments (Pillar I) and on the rural development finances (Pillar II). Direct payments have often been improperly implemented, resulting in frequent penalties charged to Greece for non-compliance with EU rules or inadequate control procedures on agricultural expenditure. During the past 7 years (2007–2013), Greece was charged with a total amount of €1.4 billion (Fig. 3.5). Even in the middle of the financial crisis, in 2010, Greece had to return €479 million. This is a result of mismanagement and poor inspection mechanisms. It is important that this problem is reduced to its absolute minimum. The absorption of rural development funds is not at all satisfactory either. Greece managed to absorb only 41% of the funds committed to the country's rural development programme 2007–2012 (EC—DG AGRI, various issues). This is extremely low, and only Bulgaria and Romania have lower absorption rates.

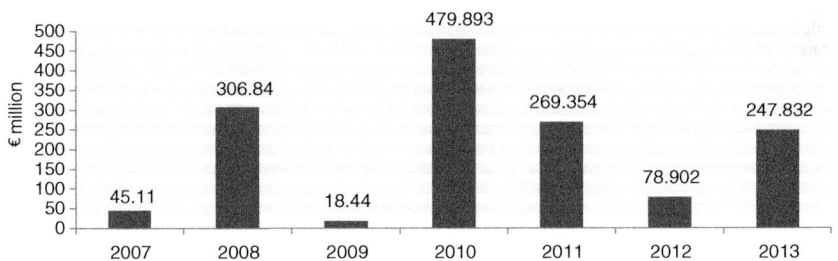

**Fig. 3.5** Returns claimed from Greece due to non-compliance or inadequate control. *Source* European Commission, own calculation

## 3.1.4 The New CAP

The 2014 was the year when a new reformed Common Agricultural Policy (CAP) will be implemented across the 28 members of the EU. The main long-term challenge that the new CAP had to face was the continuous growth of food demand globally. This demand growth, however, takes place within an economic environment of increased price volatility and in a natural environment challenged by climate change with adverse effects on productivity and the terms of trade. The rural environment in the EU is also challenged from depopulation and relocation of businesses. The rapid growth of China and India challenge both food demand and the demand for farm inputs, the prices of which also increase continuously.

After long series of consultations that started in 2010, a new CAP consensus has been reached which is based on continuous support of the producer—not the product, and support to farm practices that do not harm—but protect—the natural environment, with a balanced regional development. The main theme of the new CAP is a follow-up from the last reform, namely the joint provision of private and public goods. The new instrument of "greening" is directed to the provision of environmental public goods.

The new CAP allows more flexibility for member states in the budgeting and implementation of first pillar instruments (direct payments). Hence, each member state is allowed to choose whether they use, for example, the historical model for the calculation of payments, whether they divide the country regions based on geography, or agro-ecological practices, etc.

The new CAP advances measures that facilitate producer cooperation under both pillars. It enhances producer organizations, their associations and inter-branch organizations. It allows collective bargaining and delivery contracts, and introduces temporary exemptions for certain competition rules in periods of adverse market conditions. A high-level group for a better functioning food supply chain has been established, with the aim of improving the functioning of the chain.

On 4 June 2014, the GMRDF announced the national proposal summarized in 10 points (To Xrima 2014). Among others, the ministry's proposal (which is clearly a national choice, not necessarily dictated by CAP measures) includes a cupping of payments above €150,000, limited coupled support for some crops; special measures for small farmers and young farmers; and measures for areas under natural constraints. One key proposition is the division of the country into three zones:

I. Grasslands
II. Permanent crop areas
III. Arable land

The ministry's rationale was based on the differential productivity and land values in these areas. The zoning of the CAP is a point of controversy between the ministry and various farm organizations. PASEGES (Panhellenic Confederation of Agricultural Cooperatives) has made a comprehensive analysis of six zoning scenarios: two based on administrative and four based on agronomic criteria (PASEGES 2014). This analysis shows that a single zone scenario brings the highest benefits. This scenario favours livestock production and transfers significant amounts to pasture lands (benefits of pasture lands increase by 62%). According to PASEGES, this is also the simplest and most transparent scenario, but it does take into account agronomic criteria. It appears that the issue of the Greek proposal is not settled yet. The new leadership of the GMRDF which took office in June 2014 promised a dialogue with the stakeholders.

## 3.2 GMRDF: Greek Ministry of Rural Development and Food

The "Ministry of Agriculture, Commerce and Industry" was founded in 1910 and has undergone many structural and political changes since. In its current structure, the GMRDF was formed in 2004, and it is number 11 in the rank of ministries. An in-depth analysis of the structure, functions and the role of the GMRDF would have been very useful; however, it stretches beyond the scope of this study. We will only point to some cursory observations regarding the structure and role of the GMRDF.

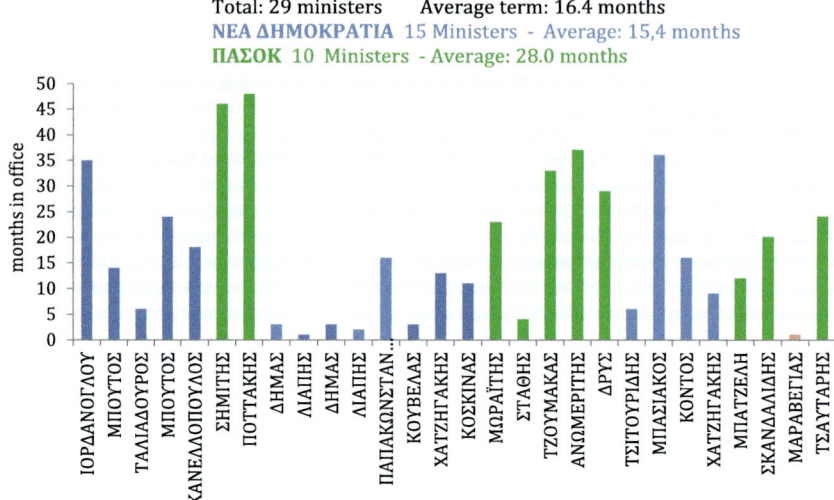

**Fig. 3.6** Ministers of agriculture, Greece: 21 November 1974–9 June 2014. *Source* GMRDF, own calculation

The political leadership in the GMRDF has been very unstable, with ministers and deputy ministers changing very often—a phenomenon not different from other ministries in Greek governments. During the period 1974–2014, there were 29 ministers with an average 16.4 months (Fig. 3.6). In the current GMRDF, in the period 2004–2014, 8 ministers have taken office with an average term of 1.2 years. Three of these ministers held office for only a few months. This pathology of frequent reshuffling of ministerial offices is endemic in Greek politics and creates a great degree of instability for such an important ministry and sector of the economy, such as agriculture, food and rural development.

## 3.2.1 The Agricultural Policy Formation and Implementation Process

It is not only the political leadership, but also the administrative structure of the GMRDF that is not functioning adequately. Political analysts find that the political process in Greece is highly centralized. The political process of decision-making in agriculture is revolving around the GMRDF, with low or absent involvement of other stakeholders, such as the regional and local governments, upstream and downstream agricultural and food industry interests, and others. The proposition to create the GIPAFC and involve it in close consultation and co-decision in some aspects, with the GMRDF, is founded on this diagnosis of absence of connection and representation in the political and administrative structure of Greek agriculture (Chatzopoulou and Lewis 2014).

The Greek decision-making system is particularly centralized. The ministerial centralized decisions with the advantageous limited circle result in an incoherent formal agricultural administrative arrangement that is considerably fragmented, decentralized and multilevel and particularly top-down hierarchical in style. The administrative structure in Greece is characterized by political scientists as "unprofessional". This term is not synonymous to "incompetent".

In Greece, the agricultural institutional and organizational settings are hierarchical, centralized, conflictual, politicized, quasi-corporatist and non-professionalized. These characteristics are reflected in the fragmented incoherent and inflexible (focussing on technicalities) administrative arrangements and practices that either fail or adapt only formally to CAP rules. Lack of professionalism combined with high politicization impedes administrative adaptation and nurtures clientelism in Greece, characterized as low Europeanization inertia (Chatzopoulou 2014).

Figure 3.7 presents the actors (nodes) that are involved in the process and their relations (ties). The network includes government actors, the regional government, the local government, the national members of the European Parliament, and national and local agricultural organizations. The different shapes correspond to the main categories of actors. The circle corresponds to sectoral organizations, the diamond to the governmental level (ministerial, administrative, regional and local) and the triangle to the EU level. The different shades distinguish the levels— black is national, light grey is regional (Prefectures), dark grey is EU, and white is local. This sociogram shows clearly that the Ministry of Rural Development and Food (MINRDF) is the central actor. The next most central actors are OPEKEPE (administration at the ministerial level),

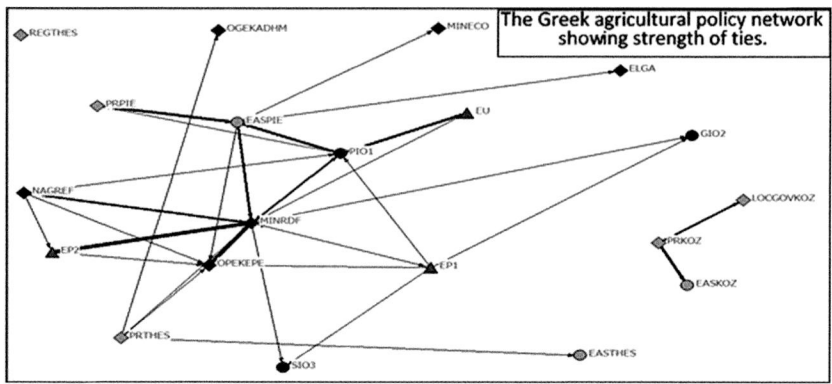

**Fig. 3.7** The Greek agricultural policy network. *Source* Chatzopoulou and Lewis (2014). Legend: PIO1, PASEGES (National Agricultural Organization), NAGREF (National Agriculture Research Foundation), EASPIE, Cooperative Union Pierias, EP1, Greek Member European Parliament (S), PRKOZ, Prefecture Kozani, EP2, Greek Member European Parliament (L), PRTHES, Prefecture Thessaloniki, PRPIE, Prefecture Pieria, ELGA (National Agricultural Insurance), GIO2, GESASE (National Agricultural Organization), SIO3, SYDASE (National Agricultural Organization), OPEKEPE (National Agricultural Administration), MINECO (Ministry of Environment), EASPTOL, Cooperative Union Ptolemaida, OGEKADHM (Organization for Agricultural Education), EU (EU Commission), EASTHES, (Cooperative Union Thessaloniki), LOCGOVKOZ (Local Government in Kozani), REGTHES (Regional Government of Macedonia)

EASPIE (cooperative union at the national level) and POI1 (sectoral organization at the national level). The mapping of interactions shows the centralization of the process at the national level. One actor, REGTHES (regional government), is an "isolate" which means it had no links to other actors.

This analysis reflects the situation at the time of the study (2007), when the regional level had no competences in agricultural policy in Greece. These are followed by the regional authorities—prefectures and the local level authorities (municipalities), the second level of the agricultural public administration in Greece. In the analysis, three large (of 52) prefectures which represent agricultural geographical areas are included: Thessaloniki, Kozani and Pieria. The prefectures interact with the local authorities (municipalities) and local cooperatives, which represent the part of the chain of the policy-making where the decisions are administered. The level of thickness of the lines indicates the frequency of interactions (the thicker, the more frequent) between actors in the process.

## 3.3 Education, Training, Extension and Research

### 3.3.1 Education and Vocational Training

Agricultural education and vocational training take place in universities, technological institutes (TEI), vocational schools of OAVETE and some independent schools.

a. *Universities*

Six universities provide agriculture and food-related education and research:

- The Agricultural University of Athens (AUA)
- Aristotle University of Thessaloniki: Faculty of Agriculture, Forestry and Natural Environment
- University of Thessaly at Volos: The School of Agriculture Sciences
- University of Ioannina: Department of Business Administration of Food and Agricultural Products
- University of Thrace: Department of Agricultural Development and Department of Forestry and Natural Resource Management
- Harokopio University, Athens: Department of Nutrition and Dietetics.

b. *TEI: Technological Education Institutes*

A large number of specialized departments at the country's 11 TEI (Technological Educational Institutes). These departments are too many to list here, but offer education in agricultural production, animal husbandry, agribusiness administration, food science and agricultural technology.

### c. *OAVETE: Organization of Agricultural Vocational Education Training and Employment*

The Organization of Agricultural Vocational Education Training and Employment—OAVETE (ΟΓΕΕΚΑ in Greek): OAVETE has now merged together with three other organizations of the GMAFRD under the single organization DEMETRA (the other three organizations are as follows: ΕΛΟΓΑ.Κ.—Greek Organization of Milk and Meat; ΕΘΙΑΓΕ. —National Agriculture Research Foundation (NAGREF); and ΟΠΕΓΕΠ (AGROCERT)—National organization of Certification). OAVETE operates seven agricultural vocational schools (AVS):

1. Averofeion AVS Larisa (agricultural machinery animal production)
2. Dairy AVS Ioannina (dairy, cheese making)
3. AVS Crete (greenhouse engineering and cultivation)
4. AVS Kalampaka (wood curving and furniture decoration)
5. AVS Nemea (viticulture, winemaking)
6. AVS Syggrou (agriculture business and landscape architecture)
7. AVS Komotini (general agriculture)

The seven agricultural vocational schools of the OAVETE are under-funded and understaffed. There is no permanent teaching staff. As such, the schools cannot perform the critical role that is absolutely necessary if Greek agriculture is to move forward. Farmers need a lifelong vocational training, in both general and specialized skills. The seven vocational schools of OAVETE cannot perform this role if they are organized and continue to function as they do today. The schools are not connected with current developments in agriculture—developments in technology, legislation, policies, or the local and international markets.

### d. *Independent Schools AFS and MAICH*

These two institutes offer high-quality agriculture and food education and research:

- **The American Farm School (AFS)**[1] in Thessaloniki is a bright example of a vocational school. The AFS is a vocational high school of very high standards, with a graduate programme as well (Perotis College). It performs research and extension in many different fields of agriculture: crops, horticulture, livestock, and in the areas of management, entrepreneurship, innovation, marketing, etc.
- **The Mediterranean Agronomic Institute of Chania (MAICH)**[2] is a graduate school, one of the four CIHEAM schools (the other three are in Bari, Italy; Saragossa, Spain; and Montpellier, France). MAICH offers education at MSc and PhD levels to a large number of international students. At the same time, MAICH is very much involved with the local economy with research and extension services. It is also a very well-recognized international research centre and manages to attract significant external funding from E.U. and national sources.

## 3.3.2 Extension and Advisory Service

The importance of extension and research in agriculture cannot be overestimated. Both contribute greatly to society's welfare, in rates that make one wonder why societies do not spend more on R&D and extension in agriculture. It has been estimated, for example, that the average annual rate of return on research in agriculture is 100%, and that of extension 85% (Alston 2010). It is important, however, to devise systems of research and extension that make these rates realized.

The Extension Service of the Ministry of Agriculture (ESMA) was established in 1951 with the aim to reorganize the agricultural sector, which was suffering the consequences of WWII and the civil war that followed (Alexopoulos et al. 2009). The ESMA system functioned with good results for more than a decade. Agricultural production increased, and producers rapidly transformed from peasants to farmers (ibid p. 178). The ESMA aged too quickly; it became bureaucratic and did not link neither with university research, nor with the private sector. The arteriosclerotic organization did not let the extension system to adjust, and to cope with the technological changes and other challenges that emerged in the 1960s and beyond.

The accession of Greece in the EU in 1981, imposed more demands on the ESMA, which instead of "going out" to the farmers, became more introverted, and even more bureaucratic, due to the new challenges of the implementation of the CAP. The changes that took place in the 1990s, mainly to decentralize the ESMA and transfer the responsibility to the prefectures did not work (Kapodistrias Plan—1997 and Kallikratis Plan —2010). The decentralization resulted in a three-level disconnected and disfunctional system of extension between the ministry, the regional and the municipal levels of government.

The OGEEKA DIMITRA was established as a semi-autonomous organization in 1997, but is in general understaffed and underfunded. Through its network of local training centres (KEGE), the DIMITRA system focussed farmers' training on EU programmes, now mainly young farmers (150 h), rural women (150 h) and short seminars (60 h) (Koutsouris 2014, p. 12). Farmers need the certificate of participation in these seminars, which, with few exceptions, tend to provide nothing more than the rubber stamp for the eligibility of funding.

The Organization of Payments and Control of Payments and Guarantees (OPEKEPE) implied the creation of a central service in Athens and its own branches at regional/subregional level which nevertheless were cut off from the Prefectural Directorates, responsible thus far for the control and payments of subsidies, grants, etc., (Koutsouris 2014).

The vacuum left by a national or even regional and topical extensional system was filled by the private input sector—mainly subsidiaries of multinational corporations that sell inputs—fertilizers, agrochemicals, veterinary supplements and drugs (Koutsouris 2014). Individual agronomists run shops, either independent or as franchises of the input suppliers and importers. This network of agronomists provides most of the extension services and advice to farmers today. Naturally, their clients are mainly professional, extensive farmers—not the small producer. Often, these agronomists are well-trained and skilled, and their services are of high quality; however, they are driven by sales of their own products, and the profitability of their business—legitimate incentives, but do not result into a concerted extension system.

It is not only technical, scientific advice that farmers need by the extension network; record keeping and eventually benchmarking of agriculture performance are also important, and are lacking under the current system. Furthermore, we found that farmers are in general uninformed about the availability of programmes, from the EU or national sources that provide investment support. Although some have access to such programmes, many remain uninformed and miss out. This is one of the reasons that Greece has such a low absorption of EU funds.

Some sporadic exceptions with good professional extension and advisory service are some producer groups which are certified under the integrated management system for agricultural production, in compliance with *AGRO 2.1* & *AGRO 2.2* standards. In such cases, agronomists are responsible for providing continuous training and technical assistance to the group's members as well as accounting and record keeping (Koutsouris 2014, p. 21).

Farmers, in general, are not satisfied with the extension system as it is today. They show mistrust to both the government-run system and the private agronomists (Alexopoulos et al. 2009; Charatsari et al. 2011). Farmers—especially young ones—are willing to pay a part of the cost for these services, if the service is provided to their standard. A totally privatized system or a system of vouchers may not be desirable (Klerkx et al. 2006). Instead, a hybrid system, both private and public, must be thought of and planned in a way that fits best the needs for Greek agriculture.

The importance of a systematic, professional extension and advisory service system cannot be emphasized enough. The absence of such a system has two related implications. First, it deprives the agri-food system —especially primary production—from one of its most important inputs: knowledge and information. Secondly, it creates a divide between (a) on one hand those farmers who can afford and are the target group, of the private agronomists with the input shops, and (b) those who cannot afford to pay for the service or—most importantly—who are not within the target group of the private agronomists/salesmen, who are not interested and are unqualified to support such groups.

Modern agriculture cannot be without continuous upgrading and information. A system of extension and advisory services that links farmers to current research and policy instruments is *condicio sine qua non* for the future of Greek agriculture. This system must eventually be owned by the stakeholders of the agri-food system. They are the immediate beneficiaries of these services, and they must take responsibility. To accomplish this, the stakeholders must organize through a system such as the pyramid GIPAFC.

Research and knowledge is an ever evolving process, and the extension service must be continuously adjusting and updating itself. A centre at the national level can provide the continuous upgrade and training of the extension experts nationwide. We propose here the creation of the Hellenic Agriculture Extension Research Centre (HAERC), which will link to national and international centres of research, the universities and other institutes of higher education and research, such as the NAGREF.

## 3.4  Research and Innovation

Greece is performing high in scientific research but very low in innovation, relative to its EU partners (EC 2014). In general research Greek scientists score high (172/200) in international scientific co-publications (EC 2014). In particular, in research related to food, Greek scientists and research institutions score high in publications in international journals (SPEED 2014).

The research infrastructure consists of a number of research centres, either government operated, or university, or independent. Some of these research centres are of very high quality and international standards. In total, there exist 51 research infrastructures in Greece (SPEED 2014). Most of these infrastructures are in Attica (29), 5 in northern Greece, 7 in Crete, 4 in Thessaly and 3 in western Greece.

The main government based agriculture and food research infrastructure is under the National Agriculture Research Foundation (ΕΘΙΑΓΕ—NAGREF) which now is under DIMITRA. The NAGREF was established in 1989 to promote agricultural research in Greece, with sporadic, but not a general success. The NAGREF used to operate 34 research units around the country and 18 research stations. Under the recent restructuring of the MAFRD, the 34 research centres will be consolidated into 10.[3]

Some institutes and research centres conduct research in the general area of agriculture, food and biotechnology at a very high level and receive international recognition:

- CERTH and IRETETH: The Centre for Research and Technology Hellas (CERTH)[4] and the Institute for Research and Technology Thessaly IRETETH.[5]
- BPI: The Benaki Phytopathological Institute.[6]

Despite some bright examples, Greece spends too little on research, only 0.5% of the GDP. What is even more critical is that research does not always lead to development and innovation. Although Greece has shown some progress, despite the downturn in 2009–2010, the country's innovation index grows, although at a lower rate than the rest of the EU (EC, Innovation Union Scoreboard 2014). Highest growth indicators are observed for community designs, community trademarks, sales share of new innovations and international scientific co-publications (EC, Innovation Union Scoreboard 2014).

What is needed is integration and coordination of research at the university, at the state and at the private and independent levels. Furthermore, research must be disseminated through an intelligent system of vocational training and extension.

# 3.5 Summary and Recommendations on Policy Governance

## 3.5.1 The CAP

The CAP is a policy that affects not only farmers, but also consumers, food processors, the tourist industry and the environment. As such, a more comprehensive governance of the policy process must be considered.

On the other hand, the formation, the negotiations and later the implementation process are a very complex matter. Individual farmers have on average little understanding of the policies, their challenges and constraints. The policy formation and implementation process, therefore, need a comprehensive representation of the stakeholders and a process that is based on consensus, rather than conflict. It has been almost an imperative that farmers will block highways, toll stations and ports every February of the last two decades. Ministers of agriculture have been sucked from office because they could not, or did not want to, deal with the farmers at the blockades. Often, policy proposals have been formed under the influence of such blockades and have been compulsive, reactionary and short-sighted. The proposed pyramid structure of the GIPAFC will lead towards the direction of broad representation and consensus building in the formation of policy.

This strong contribution of CAP funds on Greek agricultural income is useful and welcomed in the short run, but not healthy as a long-term strategy. It distorts the market, whereas a large portion of this subsidy goes to upstream (inputs) and/or downstream (distribution) industries—which have market power.

The EU financing of Greek agriculture is constituted of direct payments and rural development funds (EAFRD). The contribution of EAFRD is currently 21.7% of total CAP funds. It is important in the strategy for modulation of these payments. Greek agriculture policy should consider a further increase in the funds for rural development, directed to long-term and strategic investments. This will be more in line with the EU-15 distribution of funds, where direct payments and rural development funds are distributed evenly (48 and 47%, respectively). Regulation EU 1305/2013 allows member states to shift up to 15% of their direct payments allocation, and the funds generated from the capping or reduction of direct payments, from the first to the second pillar. Greece should develop a strategy towards this direction.

Furthermore, an additional reason and purpose of redirecting direct payment funds is to direct them towards research and development in agri-food.

## 3.5.2  Institutions and Governance

Many of the problems and drawbacks in the growth of Greek agriculture and food are of institutional nature. Their solution demands actions that are not necessarily costly financially—perhaps more costly on the political side, since decisions and break-ups with past headlocks are necessary. The political decisions and changes must be mostly on the structural level.

If agriculture and food are chosen as a driver of economic development in Greece, the stability and continuity of the institutions supporting the sector must be ensured. This can be accomplished, among other institutions, by establishing two permanent bodies of policy: a permanent secretary of food agriculture and rural development (PSAFRD) and a permanent parliamentary committee of food agriculture and rural development (PPCFRD). The positive political ramifications of these two institutions must also be taken into account. The PSAFRD being appointed and monitored by the PPCFARD will enhance a common understanding between political parties and will avoid conflicts and enhance consensus among political parties. Similarly, the PPCFRD will be more focussed on agri-food-related issues than the current Committee on Production and Trade, and will constitute a forum of consensus on critical issues of national strategy in agri-food.

e. ***PSFARD: Permanent Secretary of Agri-food and Rural Development***

It is recommended that the Greek political system establishes an institution of a Permanent Secretary of Agriculture, Food and Rural Development (PSFARD) at the GMRDF. Such a position within the GMRDF, filled by a person selected by consensus of the political parties and professional bodies, such as the GIPAFC, would establish continuity and ensure the development of long-term strategy in agriculture, food and rural development. The PSFARD will act also as a liaison between the GMRDF and the GIPAFC and, together with the Minister, will represent Greece to EU institutions.

## f. *PPCAFRD: Permanent Parliamentary Committee on Agri-food and Rural Development*

It is pertinent that the Greek parliament establishes a Permanent Parliamentary Committee of Food Agriculture and Rural Development (PPCAFRD). Currently, agricultural issues are dealt with in the 53-member Permanent Committee of Production and Trade. The PPCAFRD should be organized similarly to the 45-member European Parliament Committee on Agriculture and Rural Development (COMAGRI), but be much smaller, of course. The PPCAFRD must be a small, representative, agile committee, which will specialize on the issues, will appoint and monitor the PPCAFRD, and will establish strong, reliable and trusted channels of communication with the GMRDF and the GIPAFC. The PPCAFRD must have its own scientific support.

## g. *Mirror the EC*

To reflect on the fact that CAP is the main source of financing for the agri-food sector, the ministry (GMFRD) must reflect the organization of the EU Commission (EC), its administrative structure and processes.

## h. *Enhance Independent Research Support*

The GMFRD needs to enhance the research support so that the decisions and strategy have scientific support in both economics and other subjects. Therefore, it is essential to enhance and prioritize the role of NAGREF.

## i. *Enhance Monitoring and Control*

The monitoring bodies, such as the OPEKEPE, and the EFET must be given priority, and their role must be central in the design and implementation of policy.

j. *Work Closely with the Agri-food Industry*

The ministry GMFARD must establish closer links and work collaborate at all levels with the stakeholders of the agri-food industry. The most appropriate and effective way is to establish working committees on various issues with the GIPAFC.

## 3.5.3 Education, Training and Extension

The stakeholders of the Greek agri-food industry need to be directly involved in the funding and administration of research and extension—this can be done under the auspices of the GIPAFC.

Creation of the Hellenic Agricultural Extension Research Centre (HAERC).

Creation of a network of Agricultural Extension and Advisory Centres (AGEAC)

In the long-run, at least 1 year or equivalent full-time systematic training in one or several of the vocational schools must become a requirement for anyone who is a professional farmer. Farmers without such education should not have access to funding, or subsidies, and will have limited access to the services of the GIPAFC.

The schools must be connected with the universities and other research organizations (such as NAGREF).

The GMRDF and the research and advisory system under its auspices need high-level, well-trained staff. Not only PhD-level researchers are necessary, but the system needs more medium-level staff that is more directed to applications and extension.

The AFS and the MAICH should be studied carefully as models for vocational training, research and extension. Both schools should be given a more important role in agricultural vocational training and extension. One such role could be the training of trainers for the vocational schools.

Agricultural university faculties and TEI should develop systematic trans-disciplinary education on extension. To do so, the universities must become more involved and connected to the agricultural production and food processing and distribution systems, directly and through the GIPAFC.

Intermediate plan for the vocational schools of DIMITRA: (a) It needs to be upgraded and specialized and train future entrepreneurs in the agri-food. (b) It should receive partial funding from stakeholders for their services. (c) Stakeholders must be involved in the organization and administration of the vocational schools (VS). (d) The VS must be in close contact with the universities. (e) VS should be upgraded to continuously transfer up-to-date research to farmers.

### 3.5.4   Research

Each of the inter-professional organizations must develop their own research fund through small contributions from sales of agri-food products. Part of the research fund must be dedicated towards generic and basic research that benefits all sectors. Part of the fund must be dedicated to scholarships for graduate education and research on agri-food-related subjects.

Creation of a number of Food Innovation Parks (FIPs): Where possible, these FIPs should adapt the "triple helix" concept, where the state, university and private sector cooperate and share research and innovation projects and experiences (the addition of the regional and municipal involvement has also been considered in the quad-helix). The FIPs should not be thematic nor specialized, but should involve a multitude of research areas and scope, so that "complexity" can be created.

Food Firm Incubators (FFIs) should be developed in same location with the FIP. They should provide basic resources and financing for innovative start-ups in the agri-food sector. Young university graduates, as well as graduates from TEI and vocational and professional schools, must be provided with incentives, and financing and guarantees under the National Regional Development Program.

## Notes

1. http://afs.edu.gr/page/default.asp?id=13&la=2
2. http://www.ciheam.org/index.php/en
3. There is no final official announcement of the new NAGREF system so at the time of writing this manuscript: http://www.nagref.gr/

4. The CERTH was ranked 18th among the top European research institutions in 2012: http://ec.europa.eu/research
5. http://www.certh.gr/root.en.aspx
6. http://en.bpi.gr/

# References

Alexopoulos, G., A. Koutsouris, and I. Tzouramani. 2009. The Financing of Extension Services: A Survey Among Rural Youth in Greece. *Journal of Agricultural Education and Extension* 15 (2): 177–190.

Alston, J. M. 2010. The Benefits from Agricultural Research and Development, Innovation, and Productivity Growth. *OECD Food, Agriculture and Fisheries Papers*, No. 31, OECD Publishing.

Charatsari, C., A. Papadaki-Klavdianou, and A. Michailidis. 2011. Farmers as Consumers of Agricultural Education Services: Willingness to Pay and Spend Time. *The Journal of Agricultural Education and Extension* 17 (3): 253–266.

Chatzopoulou, S. 2014. When Do National Administrations Adapt to EU Policies? Variation in Denmark and Greece. *International Journal of Public Administration*.

Chatzopoulou, S., and J. Lewis. 2014. Analysing Networks. In *Research Methods in European Union Studies,* ed. Kennet Lyngaard, Ian Manners, and Karl Löfgren. Palgrave Studies in European Union Politics. London: Palgrave.

EC. 2014. *Innovation Union Scoreboard.* European Commission.

Karantininis, K. 2014. *Assistance to New Opportunities for Agriculture in Greece.* In Greek: Νέες Ευκαιρίες για την Γεωργία στην Ελλάδα. Task Force Greece.

Klerkx, L., K. de Grip, and C. Leeuwis. 2006. Hands Off but Strings Attached: The Contradictions of Policy-Induced Demand-Driven Agricultural Extension. *Agriculture and Human Values* 23 (2): 189–204.

Koutsouris, A. 2014. AKIS and Advisory Services in Greece. Report for the AKIS inventory WP3 of the PRO AKIS project.

PASEGES. 2014. http://www.paseges.gr/resource-api/paseges/contentObject/Koinh-Agrotikh-Politikh-meta-to-2014/content?contentDispositionType=attachment.

SPEED. 2014. SWOT Analysis for the Preparation of the Planning Period 2014–2020 (in Greek).

TO Xrima. 2014. http://www.toxrima.gr/protasi-deka-simion-gia-tin-agrotiki-anaptixi/.

# 4

# Markets

**Abstract** Farmers in Greece are faced with highly concentrated markets upstream and downstream. Land prices are still high relative to neighbouring Balkan countries. The credit market suffers from high interest rates, mainly due to the dire conditions of the bank sector, and the privatization of the Agricultural Bank of Greece. The farm inputs industry is highly oligopolistic and captures a large share of agricultural rents. The retail sector is able to charge high margins in agri-food products. Farm cooperatives are very weak, being highly leveraged and suffering from severe issues of mismanagement and corruption.

**Keywords** Credit · Retail chains · Double marginalization
Agrochemicals · Land market

© The Author(s) 2017
K. Karantininis, *A New Paradigm for Greek Agriculture*,
DOI 10.1007/978-3-319-59075-2_4

## 4.1 Land Market

The Greek land market is very thin (Swinnen and Knops 2013; Swinnen et al. 2008), mainly due to institutional factors, particularly informal institutions. More than half (51%) of cultivated land in Greece is rented (Swinnen and Knops 2013). This is lower than EU average (53%) but still a very significant amount that varies between areas and between crops. The two markets, rental and land sales, are not much in line, due mainly to institutional factors. Land is traditionally inherited from parents to children, or is transferred as dowry (προίκα) to daughter upon marriage. Although the practice of προίκα is not as prevalent in modern Greece, it has resulted in the severe partitioning of land, where the average holding is divided into five parcels.[1] The land prices and the land rents do not run in the classic Ricardian parallel fashion. Rental prices increased almost consistently since 1991 (Fig. 4.1).

Special taxation laws also favour non-sale of land, since inter-generational transfers have special tax reductions (up to 50%) in Greece. The problem of land tenure has very deep roots in Greece. The land reforms introduced in 1919–1923 initiated the recent fragmentation of farmland, by giving an average of 7 parcels to each recipient. In 1952, the re-consolidation of land (αναδασμός) was introduced in the country's constitution, and exists in today's constitution (Constitution, Article 18, paragraph 4). There were several redistributions of public-to-private lands (7000 ha, between 1996 and 2003). In the 2010

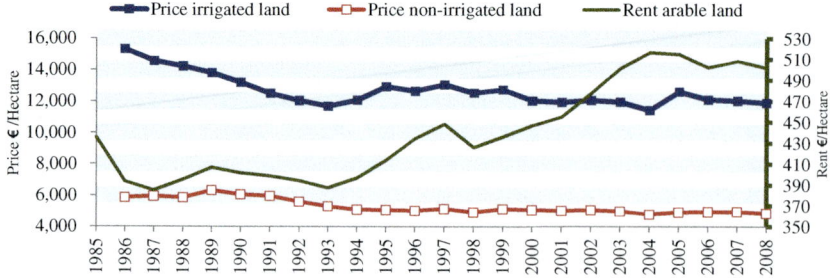

**Fig. 4.1** Price and rent for land—Greece 1985–2008. *Source* EUROSTAT, own calculation

reorganization of EUROSTAT data, Greece registered for the first time an extra 1.7 million hectares of "public lands". These are mainly grass-lands (EUROSTAT).

The dichotomous land and rental market in Greece is thinly regulated and is one of the most "liberal" compared to its EU partners (Swinnen and Knops 2013). Furthermore, it is essential that Greece up until today does not have a complete cadastre—a comprehensive, systematic and up-to-date registration of the country's real estate. The absence of a cadastre has negative consequences in the development of national agricultural policy and rural development planning.

The absence of a proper cadastre creates an institutional void with consequences that go beyond agriculture per se. For instance, there is evidence that forest fires and other wildfires are correlated with election years (Skouras and Christodoulakis 2014; NYT 2013).

The market for agricultural land in Greece is limited and very special, compared to most EU member states. Farmers usually do not sell the farm; instead, they keep it as insurance, and as their future retirement plan, by renting it to other younger, professional farmers. Most of the land for sale belongs to those who have moved to urban areas. These "urban landlords" also do not sell easily, because many of them plan to eventually move back to the rural area for their retirement. Transactions are mainly reported for plots that are located next to those of the buyer and the latter wants to increase his property and/or avoid any frictions and litigations with the neighbours. As a consequence, agricultural land sale transactions are very few and involve holdings located in plain areas and near towns. It should be noted, finally, that if part of the state-owned land is sold, then the selling price is going to drop very low.

While land prices have been consistently decreasing and land rents are increasing (Fig. 4.1), Greek land prices and rents are higher than some of its neighbouring Balkan countries, such as Bulgaria, Turkey and Romania (Figs. 4.2 and 4.3).

The lower land prices and lower rents in neighbouring countries pose a significant challenge to Greek agriculture. First, because these neigh-bouring countries can export at lower costs (besides lower land costs, they also have lower labour, and other costs of production). They can export

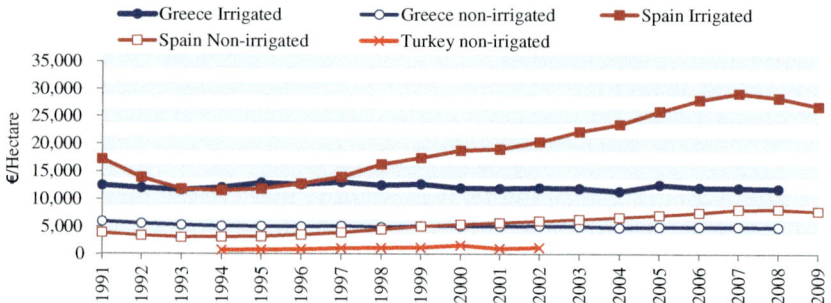

**Fig. 4.2** Land prices—Greece and selected countries. *Source* EUROSTAT, own calculation

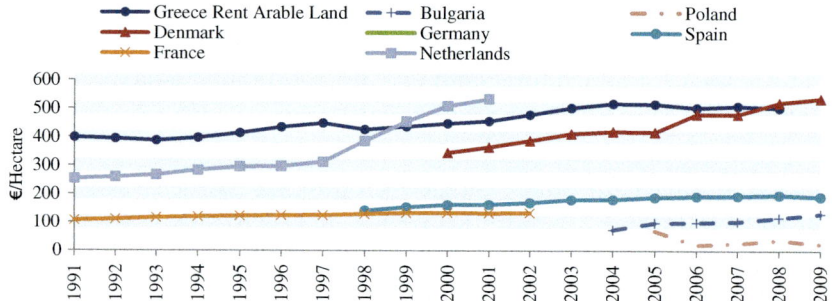

**Fig. 4.3** Land rents—Greece and selected countries. *Source* EUROSTAT, own calculation

to Greece—as they already do—and also out-compete Greek commodity products in northern European markets. Secondly, there is a potential of Greek farmers to move abroad—a movement that has already started to Bulgaria, Romania and elsewhere. The consequences of this migration of Greek "global farmers" are yet to be analysed and understood (Karantininis and Zylbersztajn 2007).

In mountainous and semi-mountainous areas, there are a limited number of land transactions, and there exist large areas of "near-abandoned" land. This is an indication of the decreased interest of potential buyers in investing in these lands for farm or non-farm activities. The low, almost zero, price of these lands is offering a tremendous potential—at least to a certain number of them.

A very good example of an integrated development which combines vertically integrated livestock production, and eco-tourism, is the mountain village Anavra, in Magnesia. With planning and collective action, this village is transformed into a model of rural development, with livestock parks, wind energy and a complete renovation of the village's infrastructure. Anavra now attracts youth and has a zero unemployment rate (http://www.anavra-zo.gr/en/the-village).

## 4.2   Credit Market

Compared to other EU countries, the agricultural sector in Greece is less exposed to credit (Swinnen and Knops 2013). The leverage of farms was only 0.6% in Greece, compared to the EU average 14.6%, and the extreme case of Denmark (49%). Also, compared to its EU northern counterparts, Greece has a very low share of outstanding agricultural loans (ibid., p. 269). On the other hand, Greece faces one of the highest interest rates (Petrick and Kloss 2013) in the EU. Also, interest charged to agriculture is significantly higher than interest rates charged to other corporations (non-financial) in the country.

Having to bear higher cost of capital, Greek farmers are therefore at a disadvantage in the credit market compared to their EU competitors. The immediate effect is on farm investments in Greece, which are among the lowest among EU countries (Petrick and Kloss 2013).

A very negative effect of the high interest rate is that farmers are more dependent to credit by upstream input firms and downstream buyers of their product. Given the non-competitive structure of these markets, the results are severe for farmers.

For agricultural production to grow, especially towards activities with high value addition, credit is necessary. Credit must come from a well-functioning credit market. The recent (2012) acquisition of, part of, the Agricultural Bank of Greece (ATE) by the Piraeus Bank is a development that could potentially have positive developments in the credit market, and consequently, in the agri-food sector in Greece. The Piraeus Bank has acquired all the network of branches of the old ATE, and with this, many of the experienced personnel with long-term knowledge of the sector has been working now under the Piraeus management. The Piraeus Bank has been developing some new instruments, such as financing contract agriculture, and others, that are flexible and may assist a certain portion of farms and certain types of agricultural production. However, the Piraeus Bank is, nearly, a monopoly in this area. Although the market is theoretically open, very few commercial banks have the know-how, and the network to enter into farm credit and into the agri-food business—especially given the current dire financial conditions of the country and the state of the banking sector as a whole. The very high interest rates are perhaps an indication of this monopolization—although this is a far more complex issue and needs careful investigation. The development of the agri-food sector in Greece needs some long-term planning and long-term investments, hence requires a long-term perspective and not a short-term view. Also, the large number of small farms poses a serious problem to any financial institution due to high transaction costs and low returns. Entry into the credit market would be very welcomed and should be encouraged, although this can hardly be a result of any policy and could only remain wishful thinking, especially under the current financial situation. Larger farms are better equipped to withstand the high cost of credit. The cooperative banks—some already operate in the agri-food industry—could potentially constitute an alternative niche in the market for agricultural credit.

For the smaller farms, however, high cost of credit can be a burden. This is further enhanced by the fact that often these farmers receive credit from input suppliers, or advances from buyers, which means higher input costs and lower product prices, which reduces profitability and long-term sustainability of the farm. To overcome this, small farms need to collect into smaller or larger entities as producers' groups or cooperatives or under other legal form. In this way, they can receive collective credit and provide mutual guarantees for loans, which may result in receiving loans under better conditions. For very small farms, micro-credit might be another alternative.

Greece has one of the lowest rates of access to venture capital in the EU (WEF 2014). This makes the credit problem even more severe, especially for new entrants—i.e. mainly young farmers and entrepreneurs. The development of innovation parks and firm incubators may create opportunities to attract outside investors.

## 4.2.1 The ATE–Piraeus Bank Hangover

Until recently, the agricultural credit in Greece was handled by the state-owned Agricultural Bank of Greece (ATE). One of the outcomes of the banking crisis and the privatization that followed was the acquisition in 2012 of the ATE by the privately owned Piraeus Bank. Today, the Piraeus Bank is almost the sole agriculture credit institution.

The transfer of the Agricultural Bank of Greece (ATE) to the Piraeus Bank was formalized with the publication of the agreement on the Government Gazette (2209/27 July 2012). With this, essentially the ATE was split and the "Good Bank" was acquired by the Piraeus Bank, whereas the "Bad Bank" was left to an appointed liquidator to handle in future. The "Bad Bank" consisted of several liabilities, including swaps and loan guarantees, as well as loans to large farm cooperatives and other large processing corporations, such as the DODONI dairy, the feed mill factory EΛBIZ. In addition to these, an amount of farmers' bad loans were also not transferred to the Piraeus Bank, but were left to the liquidator. The future of these farmers' bad loans is still unclear.

The ATE bad loans belong to three categories which started in ATE back in 2003: the Open Farmers' Loan (A.Δ.A.) and the Single Long Term Farmers' Loan (E.M.A.Δ.A.) and the "100 Settlement". The latter was a 2001 settlement, with government guarantees, the ATE forwarded with bad loans, mainly to livestock producers and island farmers. The A.Δ.A. loans were about €650 million, the E.M.A.Δ.A. about €200 million and the "100 Settlement" loans about €100 million.[2]

The total estimated bad loans to farmers that are currently in the hands of the liquidator concern about 10,000–15,000 farmers. The number of bad loans is actually double, because there are several farmers with more than one loan. There is no better information concerning these loans and the associated farms, their size, type, location, their demographic characteristics, etc. The liquidator has not as yet forwarded a solution to this problem. Both the "bad" ATE and the farmers' assets are in a lock-in situation, since they are tight-up as collaterals. A large number of these farms are small holdings. Many are livestock farms, and their investment, usually a stable, has long been written off and totally depreciated, with no market or productive value.

There is no general solution; only a careful and case-by-case realistic solution must be implemented. Farmers should pay what is possible after a careful business plan for each case.

## 4.3 Food Industry

Greek food manufacturing production followed the downturn of overall manufacturing during the financial crisis, but to a lesser extent (Fig. 4.4). It however decreased far more than the EU-15 food manufacturing. This is probably due to low income elasticity of food, relative to other products, so that demand for food is relatively unchanged. Two observations need attention: first, food production did not decrease as sharply as all other manufacturing in Greece; secondly, food manufacturing shows indications of recovery after July 2013. The latter is a very preliminary assessment; both observations, however, are indicative of the anti-cyclical nature of food production in Greece.

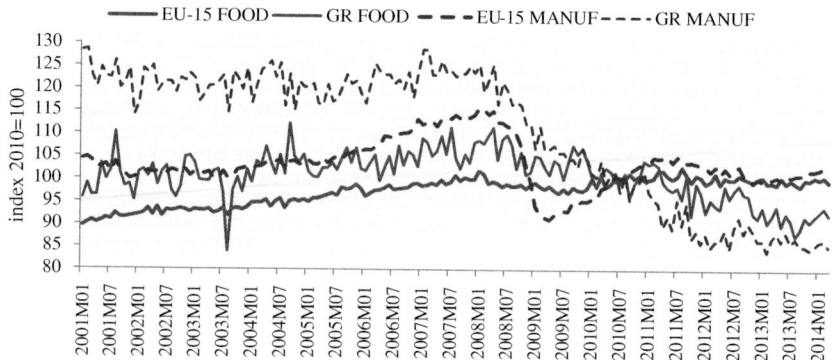

**Fig. 4.4** Production of general and food manufacturing, monthly index 2010 = 100. *Source* EUROSTAT, own calculation

Detailed studies of the Greek agri-food industry indicate high concentration and non-competitive industry structure (Rezitis and Kalantzi 2013). The overall food industry has been estimated to charge markups of over 2% during 1984–2007 (Rezitis and Kalantzi 2011)—an indication of industry concentration. Recent estimates of market power in the Greek agri-food industry corroborate with these findings. It is estimated that retailers and, to a lesser extent, food manufacturers are able to charge markups that range between 10 and 28% (Kaditi 2012).

## 4.4 Farm Inputs

Primary agricultural production is squeezed between two highly non-competitive industries—upstream inputs and downstream output. Agricultural input and output markets are not competitive, with indications of market power on both ends of the chain.

A first glimpse at input and output prices is indicative (Fig. 4.5). Right after the financial crisis (2008), agriculture output prices decreased, while the general harmonized consumer price index (HCPI) continued to increase until 2012, showing some indications of decrease in 2013. Input prices, however, continue to increase (despite a temporary drop in 2009). The agricultural output prices exhibited a slight increase in 2013; recent

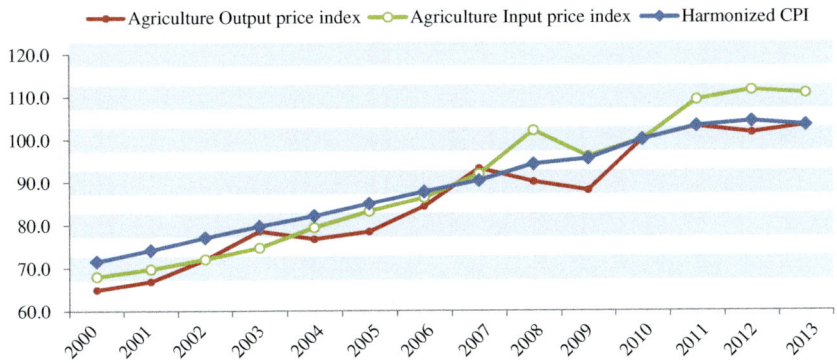

**Fig. 4.5** Agriculture input and output indices, Harmonized CPI, 2010 = 100. *Source* ELSTAT, own calculation

data, however, indicate that in the first months of 2014, output prices tend to decrease (ELSTAT). The discrepancy between input and output prices in Greek food supply chain is indicative of market power.

The total budget of the three main input items, seeds, chemicals and fertilizers, is approximately €700 million (€170 million for seeds, €170 million for chemicals and €350 million for fertilizers).[3] One should add VAT 18% and an average profit 15% on these figures. It would not be an exaggeration that the total annual budget for these three agricultural input items exceeds €1 billion annually. This is roughly 50% of the annual CAP receipts for Greece. The chemicals' budget is divided primarily between three multinational corporations.

## 4.5   Retail Sector

The retail sector is highly concentrated, like in most of the EU (Fig. 4.6). The top three retailers in Greece have a concentration ratio (CR3) 53%. This makes Greece the third highest retail concentrated market in the EU (Sweden's CR3 is 79.6%, and Ireland 61%). Supermarkets and cash and carry stores account for 90% of the total turnover of food sales in Greece. Smaller grocery shops, mini markets and small self-service stores

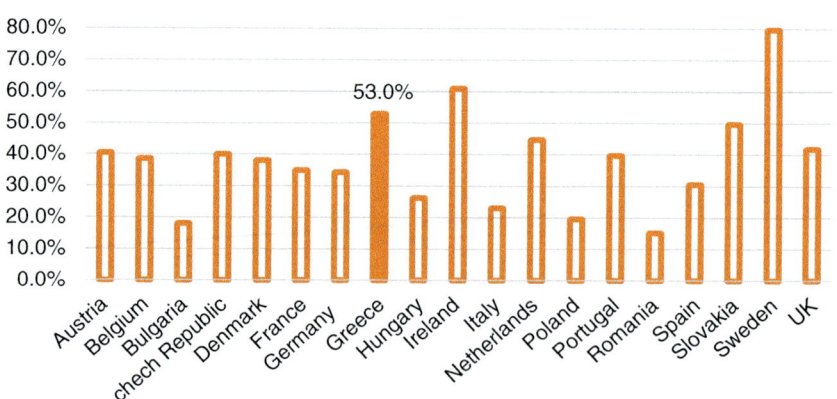

**Fig. 4.6**   CR3: Market share of top 3 retailers—Europe 2011. *Source* CIAA (2011), own construction

account for the rest 10% of sales of food (USDA 2012). Own label products have been introduced by retailers relatively recently in Greece and are estimated to account for approximately 20% of retail turnover (USDA 2012). Carrefour–Marinopoulos managed to maintain its lead over Alfa-Beta Vassilopoulos in 2012, However, the Carrefour Group decided to withdraw from the Greek market, leaving Marinopoulos Group the ownership of the Carrefour brand name in Greece (Euromonitor).

Traditional grocery stores and modern retailers have equal market share in the overall Greek grocery market (USDA 2012). Out of the €21.2 billion annual sales in 2011, €10.5 billion were handled by "modern" retailers such as supermarkets chains (€8.45 billion, or 40% of total sales). The traditional food and tobacco specialists (such as *periptera*) handle €3.6 billion, or 17% of total retail sales. Higher concentration is on the food retail, where more than 90% of the turnover of foodstuffs is through supermarket chains and cash-and-carry stores (USDA 2012). This trend is increasing, as pointed out in a recent study: "*Larger multinational players are gradually squeezing small domestic producers out, although the country's geography—with its numerous populated islands—is beneficial to small local shops and businesses*" (USDA 2012).

The market power of large retail chains is an issue not only in Greece but to most of the developed and lately of the developing world. The exercise of upstream market power by retailers is particularly severe when producers are small and not organized via cooperatives and other producers' organizations.

At a European—and later at global—level, the introduction of the EUROGAP standards in 1997 and later the GLOBALGAP in 2007 has harmonized standards on fresh produce, animal and aquaculture products procured by supermarket chains. This has improved food safety to a great extent, but also such unified strict standards increase further the market power of supermarkets. This poses a further burden on producers since it increases the power of buyers to exclude producers (Giraud-Heraud et al. 2009).

## 4.5.1  Retail Market Power and Double Marginalization

We illustrate an indicative case of exercise of oligopsony power of retailers towards producers and small traders in Greece. This illustrative example indicates a very significant case of downstream market power. The fact that the retailer is able to charge a two-part tariff and double margins is a clear indication of market power. This is a severe impediment to growth and sustainability of the agricultural sector—especially that of small scale with differentiated localized products.

This is a near "textbook case" illustrating the consequences of the concentration of retail market power. It is an indicative case of double marginalization, and hold-up by a particular wholesaler of fruits and vegetables in northern Greece, in their transaction with one of the major retail chains, which has a significant market share in northern Greece. The figures in the Table refer to actual transactions of several butches of apples that took place in 2014. It is interesting to point to the two margins that the retailer is able to extract (Fig. 4.7; Table 4.1):

**26%:** The first margin on the wholesaler's invoice is 26%, which is charged as "marketing and promotion", "returns of low quality" or "special seasonal discounts" by the retailer. There is no further indication, neither any obligation by the retail chain that they will provide any marketing service. This cost of 26% is added to the invoice, and the retailer issues an additional invoice (or more often several debit invoices that add up to 26%) to the wholesaler for the charge of the "service" equal to this margin of 26%. Often these invoices refer to "marketing services" which are under the retailer's total discretion. The seller was never provided with a proof or indication of the "marketing services" undertaken by the retailer.

**34%:** The wholesaler is able to claim the first 26% as costs in their accounting. Finally, this margin appears as part of the purchase price for the retailer who also adds on top of this the second margin: a retail profit of **34%** plus 13% VAT. The final consumer price is more than three times higher (331, 393.5, 456%) than the producer's price, whereas the total retailer's margin (including "marketing charge" plus profit) is 39.6%

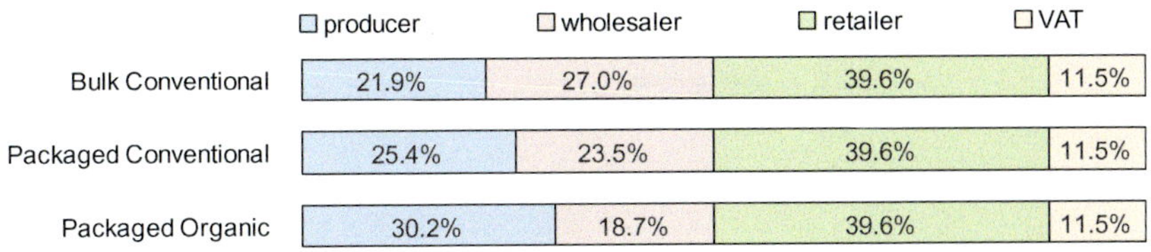

**Fig. 4.7** Share of the final price of apples. *Source* Based on calculations from Table 4.1

of the retail price or 131.3% (155.9, 180.9%) of the producer's price. The wholesaler's margin is much smaller: it is actually 61.9% (92.3, 123.1%) of the producer price or 18.7% (23.5, 27%) of the retail price.

We were able to verify these figures, and similar others with other retailers for other fruits and vegetables with some slight variations. We discovered that various retail chains have different strategies to extract an extra margin, in the form of "placement fees", merchandizing or "certification fees".

The calculations in Fig. 4.7 and Table 4.1 are only based on a case study, and we do not claim this to be a scientific representation of the procurement strategies and market power of retailers in Greece. This finding, however, corroborates with a recent analysis of competition in the Greek food chain, which finds that retailers are able to charge markups of up to 28% on food items, whereas food manufacturers can mark up at lower levels (Kaditi 2012). It is also found that foreign-owned firms (multinationals) have more market power than domestically owned firms (Kaditi 2012, p. 138).

# 4.6 Cooperatives

Cooperatives have very deep roots in Greece, dating back to early cooperatives in the nineteenth century, with the Ambelakia "Syntrofiai", which existed even before the Rochdale Pioneers in England. A law establishing cooperatives was passed in 1915 (Law 602/1915). Even the Greek constitution (article 12.4) establishes the right of individuals to form cooperatives and the obligation of the state to oversee the

**Table 4.1** An example of double marginalization. *Source* Own calculations

| | | Organic EUR/KG | % RPRICE | %prodprice | | Conventional EUR/KG | % RPRICE | %prodprice | | Bulk EUR/KG | % RPRICE | %prodprice |
|---|---|---|---|---|---|---|---|---|---|---|---|---|
| Producer price | | **1.05** | 30.20 | 100.00 | | **0.65** | 25.40 | 100.00 | | **0.65** | 21.90 | 100.00 |
| Labour | | 0.15 | 4.30 | 14.30 | | 0.15 | 5.90 | 23.10 | | 0.15 | 5.10 | 23.10 |
| Packaging | | 0.1 | 2.90 | 9.50 | | 0.1 | 3.90 | 15.40 | | 0.1 | 3.40 | 15.40 |
| Transport | | 0.1 | 2.90 | 9.50 | | 0.1 | 3.90 | 15.40 | | 0.1 | 3.40 | 15.40 |
| Transpack | | 0.05 | 1.40 | 4.80 | | 0.05 | 2.00 | 7.70 | | 0 | 0.00 | 0.00 |
| Total wholesale cost | | **0.4** | 11.50 | 38.10 | | **0.4** | 15.60 | 61.50 | | **0.35** | 11.80 | 53.80 |
| Wholesaler's profit | 0.25 | 0.25 | 7.20 | 23.80 | 0.2 | 0.2 | 7.80 | 30.80 | 0.45 | 0.45 | 15.20 | 69.20 |
| Total wholesaler's margin | | **0.65** | 18.70 | 61.90 | | **0.6** | 23.50 | 92.30 | | **0.8** | 27.00 | 123.10 |
| Marketing charge % invoice | 26% | 0.6 | 17.20 | 56.90 | 26.00% | 0.44 | 17.20 | 67.60 | 26.00% | 0.51 | 17.20 | 78.40 |
| Total invoice | | **2.3** | 66.00 | 218.80 | | **1.69** | 66.00 | 259.90 | | **1.96** | 66.00 | 301.50 |
| Retailer's profit % | 34% | 0.78 | 22.50 | 74.40 | 34.00% | 0.57 | 22.50 | 88.40 | 34.00% | 0.67 | 22.50 | 102.50 |
| VAT % | 13% | 0.4 | 11.50 | 38.10 | 13.00% | 0.29 | 11.50 | 45.30 | 13.00% | 0.34 | 11.50 | 52.50 |
| Total retailer's margin | | 1.38 | 39.60 | 131.30 | | 1.01 | 39.60 | 155.90 | | 1.18 | 39.60 | 180.90 |
| Retail P | | **3.48** | 100.00 | 331.30 | | **2.56** | 100.00 | 393.50 | | **2.97** | 100.00 | 456.50 |

cooperatives and care about their development and growth. This involvement of the state in agricultural cooperatives has been detrimental to their growth both in the 1970s and especially in the 1980s and 1990s. There were 6919 registered cooperative organizations in 1996, a number second only to Italy's in the EU. However, the state involvement was one of the key factors that led to the agricultural cooperatives' recent demise (Iliopoulos 2013). The state control on agricultural cooperatives has been very opportunistic and unfocussed and has created a great degree of institutional uncertainty to cooperatives. Characteristically, there were 946 amendments to Law 602/1915 between 1915 and 1970, approximately two amendments per month! (Iliopoulos and Valentinov 2012, p. 16). In particular, the partisan control of farmers' primary cooperatives and their federations has led to many of them being deeply in debt and close down. In fact, the management of the debt and ownership of many of these cooperative assets constitutes a major challenge today for both the farmers' cooperatives and the Piraeus Bank, the successor of the former Agricultural Bank of Greece (ATE) (Sect. 4.2.1, this book).

We do not have a clear picture of today's market share of cooperatives in the Greek agri-food chain (Fig. 4.8).

Many cooperative infrastructures, mostly built in the 1970s and especially in the 1980s, suffered from a lack of professional management and often by corruption (Netherlands Embassy in Athens Greece 2012). This is a painful truth that the Greek agricultural cooperatives need to face and deal with, sooner than later. Many of these issues were the result of badly focussed and often changing laws of agricultural cooperatives, which encouraged partisan politics in agricultural cooperatives. The 2810/2000 law for agricultural cooperatives abolishes one of the most notorious of the issues of the past by establishing a single ballot on the board elections.

The recent law 4015/2011 still leaves unsolved several problems and misgivings of agricultural cooperatives in Greece and does not allow the flexibility necessary if the Greek agricultural cooperative sector is to play a role similar to its successful counterparts in northern Europe and elsewhere. For example, the new cooperative law discourages (essentially: abolishes) the federated structure of agricultural cooperatives by not allowing unions of cooperatives (4015/2011, article 19). The role of

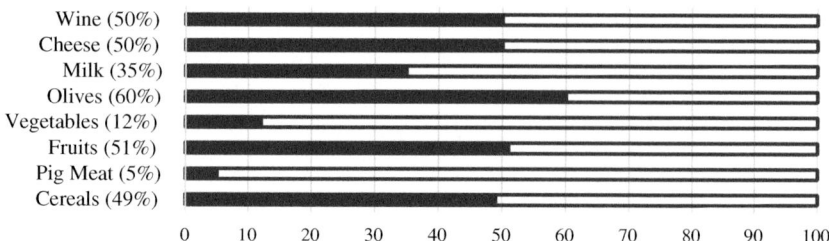

**Fig. 4.8** Market share of farmer cooperatives in the food chain: Greece 1996. *Source* Iliopoulos (2013), own construction

PASEGES as the umbrella cooperative organization is maintained, with all primary coops being direct members. Although the rationale behind such a measure is to avoid the agency problems, lack of member control and eventual corruption, at the federated level of the cooperatives, it is questionable how this new rule will lead to the growth and achievement of economies of scale by farmer cooperatives. Achieving scale economies downstream and upstream is *condicio sine qua non* for the success of agricultural cooperatives.

In order to succeed, the cooperatives need to be allowed organizational and financial flexibility. While the one-member–one-vote principle is sacred in cooperative organizations since the *Rochdale Society of Equitable Pioneers* established the cooperative principles in 1844, today's modern cooperatives use more flexible representational systems. Similarly, modern cooperatives allow outside investors and outside members on the boards of directors. A modern cooperative law must accommodate such structures.

It is well understood that a law alone cannot guarantee economic success, not only for a cooperative organization, but for any other business activity as well. It should be emphasized, for example, that some of the world's most successful cooperative sectors operate without a special cooperative law, for example in Denmark and in Ireland. However, "there are no cooperative organizations prospering without any legal rules applicable to them" (Hagen 2012). What is often even more important than legal rules and laws are a society's business ethics, social norms and social capital. These traits, however, result out of long-term processes and social evolution and are often spontaneous and path-dependent, and are not born out of conscious, planned optimizing actions. The importance of social capital is illustrated in cooperatives and social capital.

Putnam (2001) defines social capital as consisting of "social networks [among individuals] and the norms of reciprocity and trustworthiness that arise from them". He further explains that "Just as a screwdriver (physical capital) or a college education (human capital) can increase productivity (both individual and collective), so too social contacts affect the productivity of individuals and groups".

The importance of social capital in the formation, organization, success and failure of cooperatives is argued convincingly in Nilsson et al. (2012). The authors emphasize the importance of trust and reputation as the two key elements that social capital contributes to organization formation.

As an illustration, we present two case studies that initially appear unrelated, but they share a common characteristic: social capital is absent in one (mink cooperatives), whereas it is evident in the second (THESGALA). Despite the fact that the mink growers are fur processors —some of the most cosmopolitan, successful, entrepreneurial and innovative businessmen in the country, they have not succeeded to integrate and exploit the benefits of collective action. On the other hand, strong leadership, social capital and strong business drive were keys to success of the THESGALA coop.

a. *Mink growers in Kastoria*

Kastoria is a traditional fur manufacturing town. It houses some of the world's top fur establishments, some of them dating back several generations. Most of the pelts used by Kastoria fur manufacturers are imported, mostly purchased at the Copenhagen Fur Auction, and other fur auctions in Finland, Canada and elsewhere. Major export destinations of Kastoria furs are Russia, China, Europe, the USA and other international markets.

Fur animal production was not big in Kastoria, until recently. The number of mink farms grew from 5 to 70 within the past 5 years. Nine out of ten of the mink growers are also fur manufacturers. However, the astonishing fact is that the Kastoria mink growers sell their pelts through the Copenhagen Fur Auction, from where Kastoria fur manufacturers also purchase their mink pelts. It is not uncommon that a grower/manufacturer may end up buying the pelts that himself sold at the same auction!

The transaction cost by selling through the Danish auction house is about 10% of the price, or approximately €10 per pelt. Considering that there are about 1.8 million pelts produced at Kastoria annually, this is total cost exceeding €18 million annually. When asked why they prefer to sell their pelts through the Copenhagen Auction, instead of selling directly to fur manufacturers, they claimed lack of trust, payment delays and no advanced payments—elements that they can find in the Copenhagen Fur Auction. Yet, the situation would be ideal for developing an auction or a near-auction system in Kastoria, since both supply and demand for pelts are in the same location. Even more, often the same persons are suppliers and buyers of pelts.

b. *THESGALA dairy coop Larisa*

THESGALA is a newly established dairy cooperative in Larisa. Established in 2010, it now has 102 members with 58 operating farm units and handles 130 tonnes milk daily (about 10% of Greece's national production). THESGALA was a very innovative and modern organization as it introduced a network of milk vending machines—the first in Greece. It is a cooperative that is growing rapidly and recently purchased a dairy processing facility. It has established an online auction for feed procurement for its members and helped establish the growers cooperative THESGI from which the members of THESGALA procure silage and other grains. THESGALA has installed and operates a system of automatic dispensers for milk in Larisa, Thessaloniki and Athens. They have purchased and operate a dairy facility. Recently, they have launched a franchise programme for a network of small shops.

Led by a charismatic chairman, the THESGALA cooperative members are dynamic, progressive and very engaged. They have concrete growth plans for the future of their cooperative and their personal dairy business.

These dairy producers have realized a window of opportunity through cooperation.

## 4.6.1 Cooperative Restructuring, Conversion, Demutualization

A more critical and acute problem for Greek farm cooperatives today has to do with a number of cooperatives and federated structures that have been transferred to the "Bad Bank" regime and are in the hands of the liquidator. The process is stagnating. Many of the infrastructures remain unutilized or they are operating and ownership state that is unclear and not transparent. This stalemate does not help the future of agri-food. Looking backwards in a strict book accounting manner is not a choice. Instead, a forward-looking strategy should be adopted. The legal issues under the common and criminal law must be dealt with within the appropriate institutions and political processes, whereas the business part needs to be separated and needs to move forward. It is important that this issue is viewed in a broader context than its strict financial dimension and must be forward—instead of backward—looking. Restructuring and conversion must be options to be considered.

Organizational business failures are not unique to Greece or to cooperative organizations. Business fail and bankrupt around the world every day. Cooperative organizations are not immune to failure and should not be singled out for this reason. Many cooperative organizations around the world convert to IOFs or restructure and demutualize. The reasons for cooperative conversions are many, but could be categorized into three groups: managerial incompetence and hubris; need for capital; and adjustment to needs of new business environments (Fulton and Hueth 2009). Each case is unique, and Greek farm cooperatives have their own share of institutional failures, mismanagement and hubris.

A demutualization should be considered. A strategic evaluation of a case-by-case is pertinent now. The process should involve banks (both commercial and cooperative banks), producers' groups, producers' organizations and individual producers, as well as IOFs, individual investors, venture capitalists and entrepreneurs. Also, foreign investors must not be excluded. The strategic importance of the issue must be recognized and needs to be handled in a process that involves—but it is not limited to—the liquidator.

The restructuring and opening of cooperatives to more flexible schemes should not only be limited to those cooperatives under liquidation. Healthy existing cooperatives and new cooperatives should be allowed all the organizational flexibility necessary in order to succeed—as long as they maintain their fundamental principles of cooperation and producer orientation.

## 4.7 Summary and Recommendations

### 4.7.1 Land Market

Neighbour countries with factors of production at lower prices than those in Greece, is a further reason to argue that Greece should not enter into a competition of low-cost commodity products. Since competition can come from neighbouring Balkan countries, as well as from unexpected entrants around the world (such as from Latin America, Africa and even Australasia), Greece should build its competitive advantage around high-quality, highly differentiated products, produced by well-trained farmers and distributed through an integrated sustainable supply chain.

### 4.7.2 Credit Market

The credit market is highly oligopolistic, and in addition to the dire conditions with the financial crisis, there is a high burden on capital costs on Greek farms.

The issue of the acquisition of the ATE by the Piraeus Bank needs to come to an end. The liquidator actually takes ownership and puts the farms to sale. This is highly problematic, since very often, not only the farm, but also the house and other assets are leveraged. Also, putting at once a large amount of land for sale will drive land prices down, and the final benefit to the liquidator (i.e. the government) will be much less than anticipated.

It has also been proposed—and tried in a limited number of cases—that the "bad" ATE takes ownership of the farm and charges a rent to the farmer. This can be devastating to a farmer. Land is not only an asset, but also a means of production and has much more than financial value to a farmer. Land is most often inherited from generation to generation and is associated with a farmer's societal and economic status. Instead, the Piraeus Bank (or potentially, any other bank or several com-mercial banks) could renegotiate and buy back these loans at a negotiated discount, and these farmers could start from a better position.

We highly recommend a cooperative scenario, where farmers are encouraged, assisted and financed, to voluntarily collect resources and achieve economies of scale by forming a cooperative or even an IOF. In this way, the potential pay-off of the loan increases, as well as the economic viability of the farmer. At the same time, this may increase the overall quality of production and sales of agricultural products, since they will be organized, professionally and in coordination. Potentially, a branch, or a joint venture between the GIPAFC and the Piraeus Bank and perhaps in a P-P partnership with the government who is an immediate stakeholder through the liquidator, may provide advice, management and financing of small and larger such cooperative or IOF units.

### 4.7.3 Input and Output Markets—Cooperatives and the New Paradigm

Both input and output markets are highly oligopolistic. A small number of agrochemical multinationals control the input markets, while retailers often are able to extract high rents from upstream.

It is important that the agri-food sector organizes under a new paradigm which involves both private and public sectors and links agri-food to other industries. The core of this new paradigm is the Greek Inter-Professional Agriculture and Food Council (GIPAFC).

The GIPAFC (presented in more detail in Chap. 6) is the pyramid organization of the stakeholders of the agri-food chain. It will provide services and lobbying to its members.

Collective action by producers is key. When farmers are organized into producer groups, cooperatives or other collective legal forms, they gain market power vis-à-vis the retailers and input sellers.

There is a need to maintain and expand, where necessary, processing facility for low-quality produce. In particular, in horticultural products, such as fruits and vegetables, there is a variety of quality. It is very difficult for farmers to dispose of low-quality products, and often they are forced to sell all their crop at a lower average price. Most often, this price–quality relation is not reflected at the retail level, since retailers are able to mix qualities or use other methods. If there was another market for lower quality products, such as juice for fruits, then the higher quality products could be sold at a much higher price for producers. It is therefore necessary that such facility is introduced at the necessary large scale to absorb and process low-quality horticultural products. An investor-owned firm, or more preferably, a farmer-owned cooperative, or a joint venture between farmers' cooperative(s) and IOFs, that would invest in a region-wide or nationwide large-scale facility for processing lower quality fruits and vegetables would increase competition in the market in favour of producers.

Organization of farmers' markets may reduce, to some extent, the retailer's market power and offer an outlet for small, high-quality producers. A proper, competitive, controlled system of farmers' markets increases competition and relaxes retailers' market power.

It is pertinent to enhance the role of the Competition Committee and expand its role to both input and output markets.

## Notes

1. Unconfirmed figure.
2. Piraeus Bank, private communication.
3. Based on private communication with industry experts, the figures could not be verified, but were cross-checked and confirmed by various independent sources.

# References

CIAA. 2011. Data and Trends in the European Food and Drink Industry. http://www.fooddrinkeurope.eu.

Fulton, M.E., and Hueth, B. 2009. *Cooperative Conversions, Failures and Restructurings: Case Studies and Lessons from U.S. and Canadian Agriculture.* Saskatoon, Sask: Center for the Study of Cooperatives, University of Saskatchewan.

Giraud-Heraud, E., C. Grazia, and A. Hammoudi. 2009. Agrifood Safety Standards, Market Power, and Consumer Misperceptions. *Journal of Food Products* 16 (1): 92–128.

Henry, Hagen. 2012. *Guidelines for Cooperative Legislation.* Geneva: International Labour Organization.

Iliopoulos, C. 2013. Support for Farmers' Cooperatives: Country Report Greece European Commission. http://ec.europa.eu/agriculture/external-studies/support-farmers-coop_en.htm.

Iliopoulos, C., and V. Valentinov. 2012. Opportunism in Agricultural Cooperatives in Greece. *Outlook on Agriculture* 41 (1): 15–19.

Kaditi, E. 2012. *Analysis of the Greek Food Chain.* Athens: Centre of Planning and Economic Research (ΚΕΠΕ).

Karantininis, K., and D. Zylbersztajn. 2007. The global farmer: typology, institutions and organisation. *Journal on Chain and Network Science* 7 (1): 71–83.

Netherlands Embassy in Athens Greece. 2012. Developments in the Greek Horticultural Sector.

Nilsson, J., G.L.H. Svendson, and G.T. Svendson. 2012. Are Large and Complex Agricultural Cooperatives Losing their Social Capital? *Agribusiness* 28 (2): 187–204.

NYT. 2013. Who Owns This Land? In *Greece, Who Knows?*

Petrick, M., and M. Kloss. 2013. Exposure of EU Farmers to the Financial Crisis. *Choices* 28 (2).

Putnam, R.D. 2001. *Bowling Alone: The Collapse and Revival of American Community.* New York: Simon and Schuster.

Rezitis, Anthony N., and Kalantzi Maria. 2011. A. Investigating market structure of the Greek manufacturing industry: A Hall-Roeger approach. *Atlantic Economic Journal* 39 (4): 383–400.

Rezitis, Anthony N., and Maria A. Kalantzi. 2013. Measuring the Degree of Market Power in the Greek Manufacturing Industry. *International Review of Applied Economics* 27 (3): 339–359.

Skouras, S., and N. Christodoulakis. 2014. Electoral misgovernance cycles: evidence from wildfires and tax evasion in Greece. *Public Choice* 159 (3–4): 533–559.

Swinnen, J.F.M., and L. Knops. 2013. *Land, Labour, and Capital Markets in European Agriculture: Diversity Under a Common Policy*. Brussels: Centre for European Policy Studies (CEPS).

Swinnen, J.F.M., P. Ciaian, and D.A. Kancs. 2008. Study on the Functioning of Land Markets in the EU Member States under the Influence of Measures applied under the Common Agricultural Policy. Unpublished Report to the European Commission. Brussels: Centre for European Policy Studies.

USDA. 2012. GAIN Report. Greece: Retail Foods. GR 1209.

WEF. 2014. Enhancing Europe's Competitiveness. Fostering Innovation-driven Entrepreneurship in Europe. World Economic Forum (WEF).

# 5

# Fruits and Vegetables, Aquaculture, Olive Oil, Organics, PDO, PGI

**Abstract** F&V, olive oil and aquaculture are three very dynamic and export-oriented agri-food chains. F&V exhibit a high seasonality, which is due to natural conditions as well as high demand during the summer months due to tourism. Stone fruits exports are depended on eastern Europe and Russian market, and have suffered greatly from the Russian embargo. Greece is a major producer of aquaculture in the EU, but the industry undergoes a credit crunch. Olive oil has a great potential, but currently is exported bulk and at low prices. The examples exemplify the need for a new paradigm which will take advantage of the complexity of the Greek agri-food industry.

**Keywords** Greek fruits and vegetables · Greek aquaculture · Greek olive oil · Complexity in agriculture · Isomorphism

© The Author(s) 2017
K. Karantininis, *A New Paradigm for Greek Agriculture*,
DOI 10.1007/978-3-319-59075-2_5

## 5.1   Introduction

It is not possible within the limitations of this book to cover the entire agri-food industry of Greece. It is very diverse and would take time and space that are beyond the limits of this work. The three agri-food chains we present here are as follows: (a) fruits and vegetables—a dynamic industry important for both domestic and export markets; (b) olive oil—a traditional industry, with potential to grow; and (c) aquaculture—an industrialized, export-oriented and highly competitive industry. We also provide a short overview of organics, PDO and PGI.

## 5.2   Fruits and Vegetables (F&V)

After the EU accession in 1982, we observe a decrease in traditional crops, such as tobacco and fruit trees. With the exception of olives, most tree crops decreased—at least according to available data (Eurostat). According to the most recent data (ELSTAT 2014), fruit trees decreased further during the period of the financial crisis. All fruit tree categories were reduced between 2009 and 2012 (Fig. 5.1). The total area under fruit trees decreased by 10820 ha (−10%), and the number of holdings decreased by 3012 holdings or −1.9% in the period 2009–2012. The average size of orchard decreased from 0.66 ha in 2009 to 0.61 ha in 2012; in other words, the average orchard became 8.3% smaller (Fig. 5.2). The decline in F&V cultivation and number of farms is a general trend in EU-27 countries where the total area under F&V decreased by −6% between 2003 and 2010 and the number of holdings by −39.1% (−6.3% in ha and −26.3% in holdings in EU-15). The average size increased by 1.9 ha in EU-27 and by 2.0 ha in EU-15 (EC 2014a, b). This is in contrast to the Greek case, where the average size of farms decreased further, an indication that it is not the smaller, but rather the medium and large farms that exit F&V production in Greece.

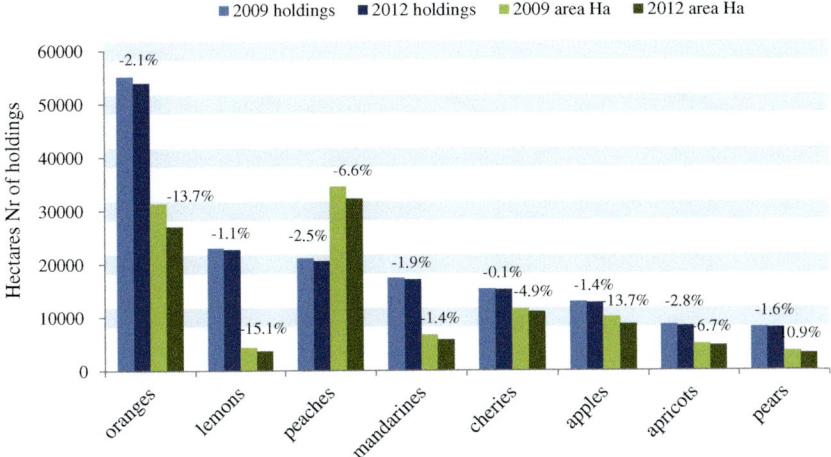

**Fig. 5.1** Fruit trees, Greece: 2009–2012. *Source* ELSTAT, own calculation

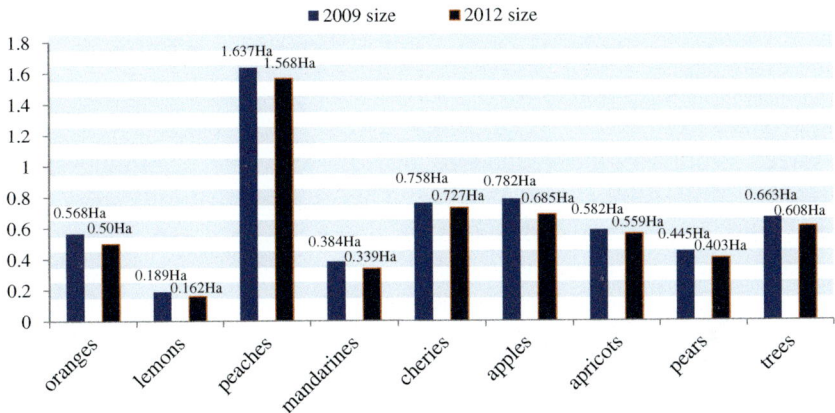

**Fig. 5.2** Average orchard size for selected fruit trees, 2009–2012. *Source* ELSTAT, own calculation

## 5.2.1 Fruits and Vegetables Trade

Fruits and vegetables constitute an important part of Greek agricultural production and exports. Fruits constitute 19.6% of total agricultural goods output, and vegetables constitute 16.3% in 2013 (DG-AGRI Fact

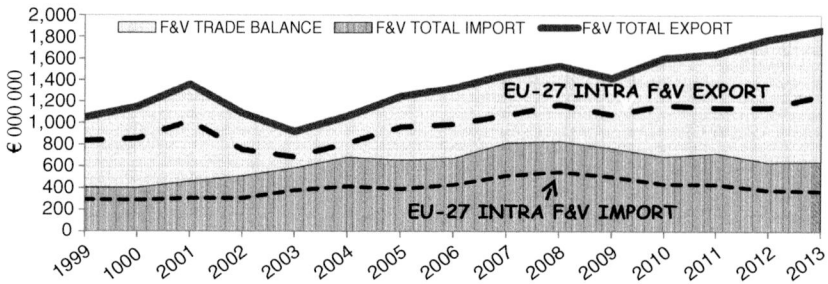

**Fig. 5.3** Fruit and vegetable trade, Greece: 1999–2013. Source EUROSTAT, own calculation

Sheets). The international trade of F&V from Greece exhibits a growing trend (Fig. 5.3). Although we observed an increase in imports of F&V until before the financial crisis (2008), imports tend to decrease after that period, while exports increase. In 2013, total exports of F&V were €1.857 billion and imports were €0.643 billion, which left a positive balance for Greece of €1.214 billion. Most of F&V is with EU countries. Of the total F&V exports, 67% are to EU-27 countries, whereas 57% of imports are from EU-27 (Fig. 5.3).

Greek exports of F&V tend to increase after the beginning of the crisis, i.e. after 2008 (EUROSTAT). This is a positive indication. This trend needs to continue and be enhanced in future. Still Greece imports a significant amount of F&V—why does Greece import fruits and vegetables? Greece imports F&V mainly to cover local consumption during off-season months. Even during the high-season tourist months—June, July and August—Greek exports of F&V are at peak (Fig. 5.4). A closer look at some key F&V categories offers a clearer picture.

Export of fresh tomatoes, in 2013, for example, peaked in April at €2.7 million (Fig. 5.5). Exports decreased after that to almost zero in June and increased slightly thereafter. Imports, on the other hand, were moderate during spring and summer months and peaked in September and October (€0.75 million and €1.2 million, respectively). Similar pattern is observed during most years. The balance of tomato trade is mostly negative, although during the last 3–4 years, it exhibits a positive

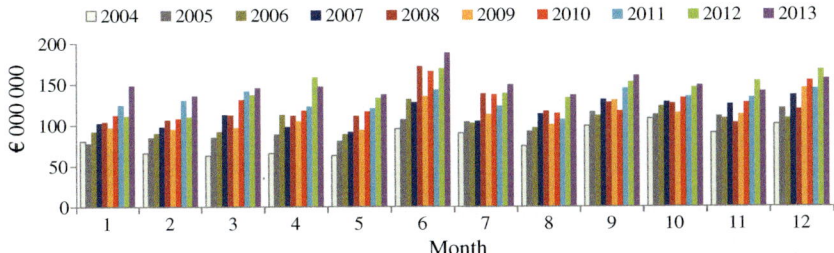

**Fig. 5.4** Fruit and vegetable trade monthly balance, Greece: 2004–2013. *Source* EUROSTAT, own calculation

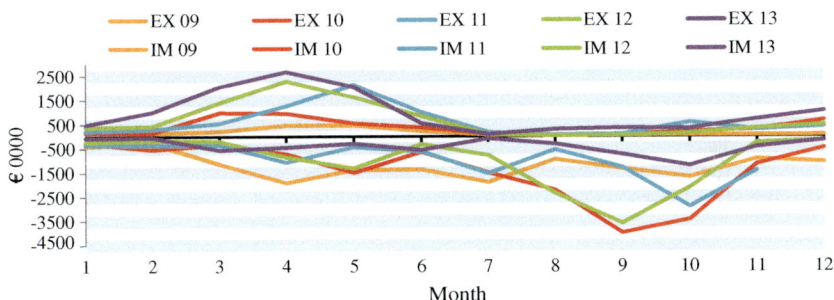

**Fig. 5.5** Tomato monthly trade, Greece: 2009–2013. *Source* EUROSTAT, own calculation

trend, especially during spring months (Fig. 5.4). Apples show a similar pattern (Fig. 5.6).

Until recently, Greece has been deficient in apples (Fig. 5.6). The balance of trade is highly negative, especially during the summer months, up until 2011. After 2012, however, the balance of trade in apples has been positive overall (€8.8 million in 2012 and €18.5 million in 2013).

## 5.2.2 Peaches

Peaches and other stone fruits have been traditional export leaders in the Greek F&V trade. However, the first years after the country's EU accession in 1981 were characterized by what has notoriously termed

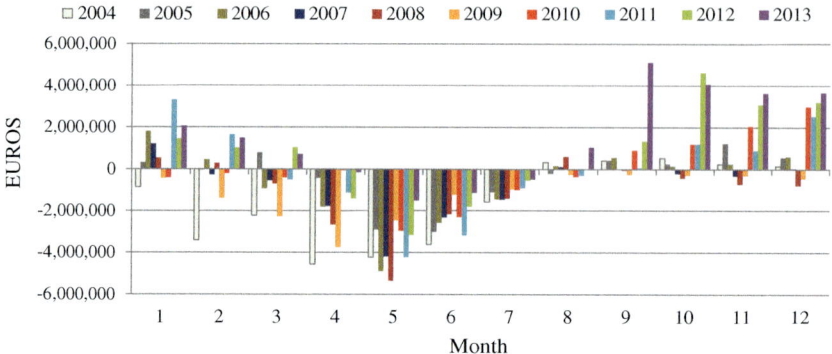

**Fig. 5.6** Apples monthly trade balance, Greece: 1999–2013. *Source* EUROSTAT, own calculation

"χωματερή"—where mainly peaches (as well as other fruits and fresh produce) were withdrawn from consumption and were buried in specially created dumps (χωματερή). They were of course paid by CAP subsidies especially put in place to protect market prices in periods of excess supply. Peaches were grown with the specific purpose to be dumped, and production of peaches increased competing with other fruits, such as cherries, pears and others. Fifty producer groups in Imathia prefecture alone were formed especially to deal with peach dumping (Vlachos and Karanikolas 2013).

The export of peaches and other stone fruits naturally exhibits a seasonality (Eurostat). In June 2013, exports of stone fruits were €61 million, whereas total annual exports of stone fruits for 2013 were €140 million.

The stone fruit exports (mainly peaches) are a good example of the form of trade diversion that occurred in some items of Greek agri-food exports. Traditionally, Germany had been the main importer of Greek peaches. More than 1/3 of the value of exported stone fruits were going to Germany, whereas this share is now taken by Russia (Fig. 5.7). Recently, Russia and other CEE countries, such as Ukraine, Romania and others, have been the main destinations of Greek stone fruit exports (Eurostat). There is a significant qualitative difference in the export destinations between CEE countries and EU countries, since the former

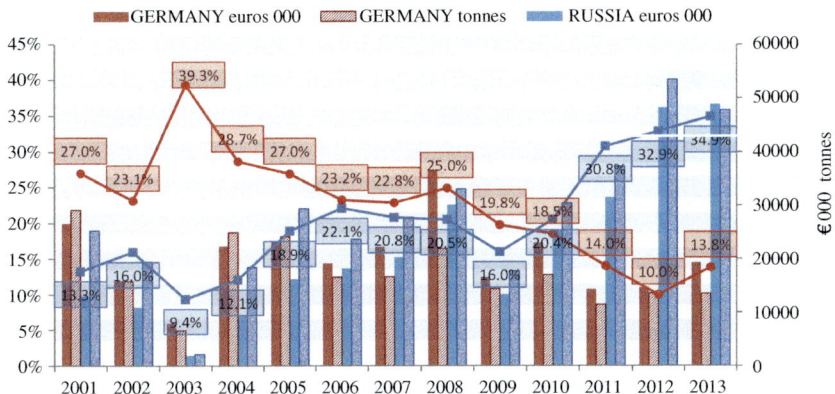

**Fig. 5.7** Stone fruits export diversion: 2001–2013. *Source* EUROSTAT, own calculation

appear to be of lower value. The average FOB price of stone fruits exported to Russia was €1025/tonne, whereas those exported to Germany were 40% higher, at €1430/tonne (Fig. 5.8). Similarly, the average value of exported stone fruits to Romania was €705/tonne, €506/tonne and €459/tonne, for Ukraine, Romania and Bulgaria, respectively. The average price to the Netherlands was €2495/tonne and to the UK was €2041/tonne (Fig. 5.8).

The stone fruits have a great potential in Greece. One avenue is through further organization and certification via the integrated production programmes of AGROCERT and other private certification agencies. Integrated facilities that process excess quantities of fruits, especially those of lower quality, are necessary to clear the market and maintain high prices for the high-quality produce. Diversification of exports to countries other than Russia would improve Greece's long-term strategic position in this trade.

## 5.2.3 Cooperatives in the F&V Sector

Indicative of the trend towards CEE exports and the implied profitability is the recent acquisition of the fruit packaging plant of the Veria cooperative union by a consortium of Greek and Russian investors. The

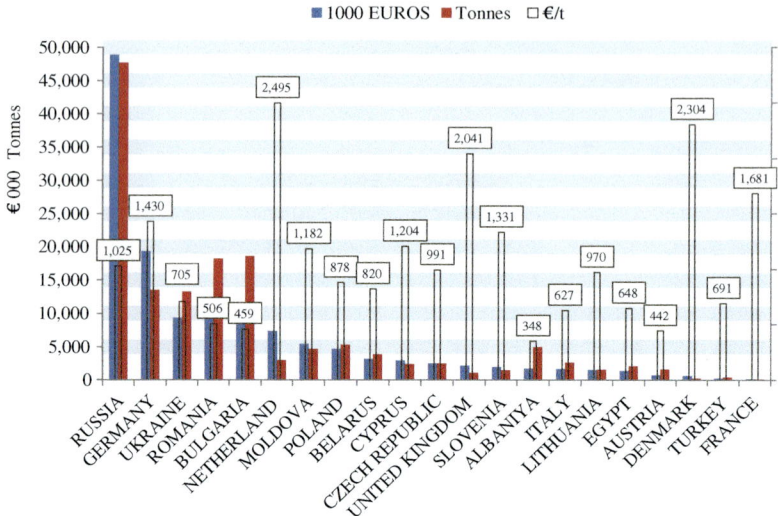

**Fig. 5.8** Stone fruits export by destination: 2013. *Source* EUROSTAT, own calculation

cooperative fruit packaging plant was heavily in debt to the former ATE bank and was transferred in the hands of the liquidator after the acquisition of ATE by the Piraeus Bank in 2012. Unfortunately, many such cooperative plants are under similar condition, heavily indebted and under liquidation.

It is estimated that the market share of producer organizations (cooperatives and unions of cooperatives) is approximately 51% in fruits and 12% in the vegetables (Fig. 4.8). There exist, however, some powerful cooperatives functioning in the F&V sector, such as VENUS in Veroia, the ASEPOP Velventos (peaches), KIKU Hellas (apples), Kirros (kiwis and other fruits), Zagorin (apples) and others (Embassy of Netherlands 2012).

Corruption, lack of professionalism and poor management were the main reasons for failure. As a result, the idea of producer organizations still brings bad memories to mind and is often not favourably perceived by most small farmers, especially the most competitive ones. (Embassy of Netherlands 2012, p. 16)

This statement encapsulates much of the problems and the general opinion on cooperatives in Greece. We are discussing some of the more general issues of cooperatives in Greece in Sect. 4.6. Cooperatives and cooperative unions, and federations of unions of cooperatives, played a very significant role in the development of the F&V sector in the past. This is not the case today. Part of the reason is captured in the quotation above; the other reason is that CAP subsidies have to some extent attracted farmers away from tree crops and towards more extensive production, such as arable crops (Fig. 3.1).

The cooperative organizations are, however, a powerful instrument when used properly. This has been recognized and promoted at the EU level in both the 1996 and 2007 reforms of the CAP. Not all countries have participated, and it is not reported whether Greece participated in the programme and to what extent (EC 2014a, b). This is perhaps due to the fact that the programme requires national contribution, which was difficult during the financial crisis (EC 2014a).

## 5.3  Aquaculture

Aquaculture has been a success story in Greece, until recently. Greek production grew in the 1990s and 2000s, mainly with marine cultures (Fig. 5.9). Greece is one of the top five aquaculture producers in the EU, the other four being France, Italy, Spain and the UK (Fig. 5.10).

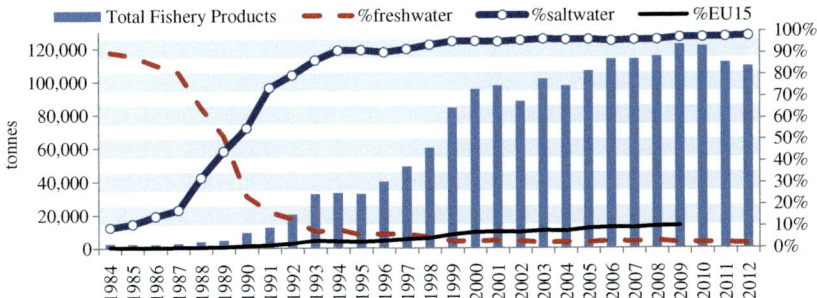

**Fig. 5.9**  Aquaculture in Greece: 1984–2012. *Source* EUROSTAT, own calculation

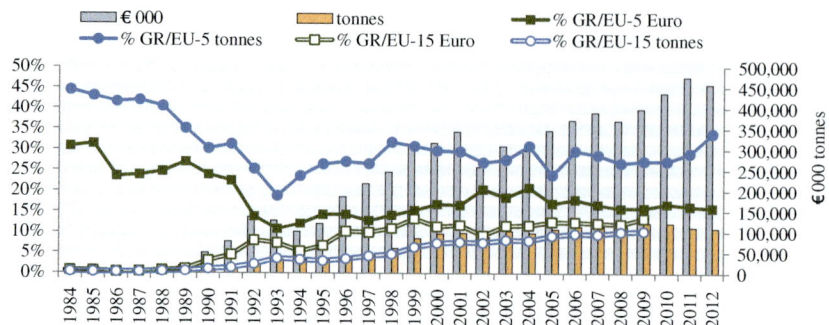

**Fig. 5.10** Aquaculture share of Greece: 1984–2012. *Source* EUROSTAT, own calculation

Turkey, a non-EU member, has become a major competitor, and has surpassed Greece after 2004, in both volume and value of total aquaculture production (Eurostat). Greek aquaculture production is, however, high in volume, but not as much as in value, because it is exported mainly as bulk, and not packaged ready for supermarket sales (Fig. 5.10).

The share of Greek aquaculture production, among the EU-5 top European aquaculture producers, is close to 30% in volume, but less than 20% in value (Fig. 5.10). When compared to total EU-15, share of value is higher than share of volume, indicating that Greek production has a slightly higher value added than overall EU, but not as much as the top-5 European producers (Figs. 5.11 and 5.12).

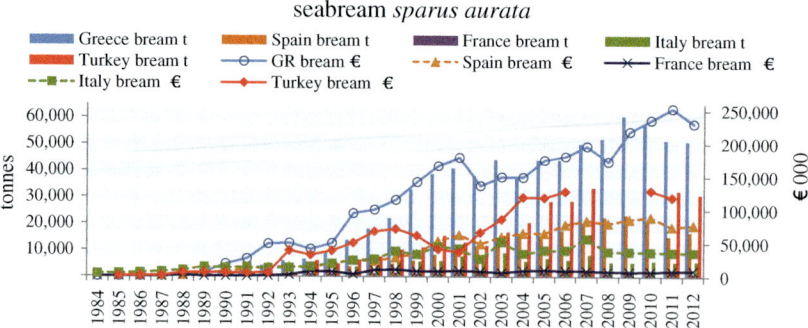

**Fig. 5.11** Production of sea bream by top-5 European producers: 1984–2012. *Source* EUROSTAT, own calculation

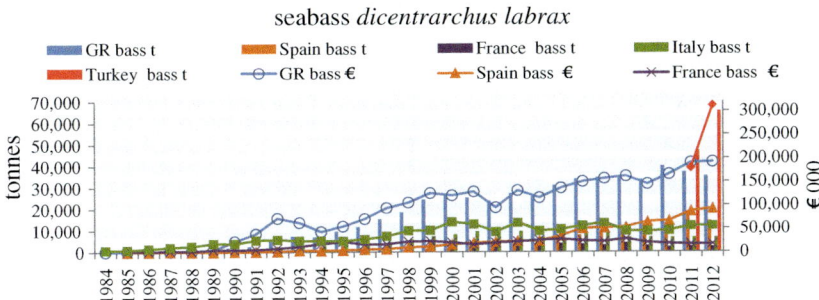

**Fig. 5.12**  Production of sea bass by top-5 European producers: 1984–2012. *Source* EUROSTAT, own calculation

The top two species produced by aquaculture farms in Greece are sea bass (*dicentrarchuslabrax*) and sea bream (*sparusaurata*). Greece is by far the largest EU producer in volume and value, of these two species (Figs. 5.11 and 5.12). Turkey, however, has appeared as a major competitor and surpassed Greece in the production of sea bass (Fig. 5.12).

It is evident also in these two species where Greece dominates the EU market, that value addition is not satisfactory. In sea bream, Greek value share is around 60% of the EU-4 average, whereas volume share is higher, more than 65% (Fig. 5.13). Similarly, with sea bass, volume share of Greece lags about 5–10% points of that of volume, among the top EU-4 producers (Fig. 5.14). These are further indications that Greek

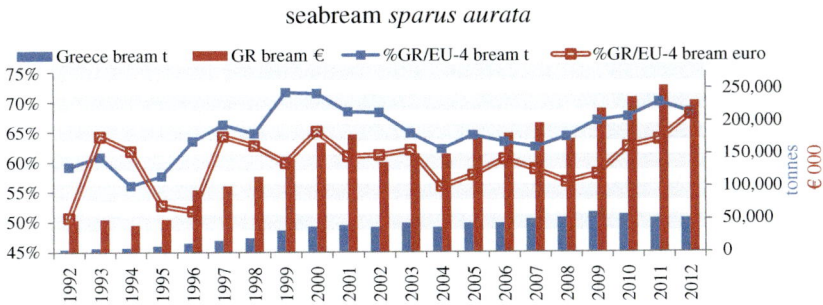

**Fig. 5.13**  Production of sea bream and share of Greece in the EU-4: 1984–2012. *Source* EUROSTAT, own calculation

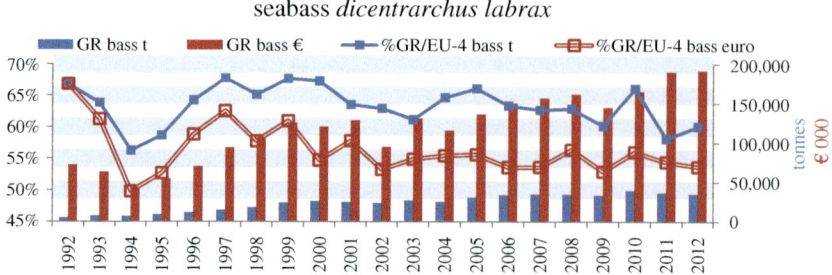

**Fig. 5.14** Production of sea bass and share of Greece in the EU-4: 1984–2012. *Source* EUROSTAT, own calculation

aquaculture production lacks the value-added activities, compared to its EU competitors.

In the 1980s, there were 10 aquaculture firms in Greece, whereas there were more than 1100 registered in 2011 (Reuters 2013). Many of the large firms are registered in the Athens stock exchange, and they boast on the top-100 list of agri-food firms in Greece (firms 9, 14, 18, 39, 45, 77, 91, http://www.inr.gr). The firm DIAS (14 in the list) filed for bankruptcy in 2013. Many aquaculture firms were hit hard by the financial crisis. The main reason was credit shortage. Many of these firms were highly leveraged during periods of cheap credit. The decrease in demand due to the financial crisis, a drop in prices and fierce competition by Turkey, especially in the two main species that Greece was market leader (sea bass and sea bream), accumulated the pressure. Today, many major aquaculture firms are pursuing mergers.

As expected, the crisis has also affected the aqua-feed industry, although not to the same extent. The aquaculture feed manufacturers increased prices by up to 15% in 2012 and manage their finances by reducing depreciation by 20% (http://www.inr.gr). The top-5 aquaculture feed manufacturers exhibited a 5% increase in gross profits between 2011 and 2012. This is an indication (still—not proof) of the presence of market power in the aqua-feed industry.

There were good reasons why aquaculture grew to a commercial success in Greece. The country has some competitive advantages in aquaculture: availability of suitable environment for marine aquaculture;

skilled labour; and knowledge of production technology (Jordi et al. 2013). The main cost items are raw materials and feed which account for 50% (ibid.). Labour cost is about 17% of total costs of production of sea bass and sea bream—the two main species of Greek aquaculture. Given the need for specialized and skilled labour, this is not a key item for cost savings for aquaculture.

The aquaculture industry, however, suffers from the credit crunch, as well as from other institutional constraints. For example, *"Due to the lack of spatial planning for aquaculture, subsidies of the European Fisheries Fund (EFF) were not granted for 2010 and 2011. During 2011, a special framework for aquaculture spatial planning came into force in Greece and the EFF grant approvals are expected during 2013"* (Jordi et al. 2013, p. 199).

To our knowledge, this has not yet been accomplished. It is urgent that the MFRD acts on this issue very quickly. Furthermore, applications for the development of aquaculture projects take long time, are very complex and involve a multitude of agencies. There is a need to get consensus by 6 major agencies: Ministry of Agricultural Development and Food, Ministry of the Environment, Navy, Archaeology, Ministry of the Merchant marine and Greek Tourism Organization (FAO).

Besides immediate need for credit, and simplification of licensing rules, there is need for long-term planning and integrating aquaculture into the development of coastal areas, in a way that does not conflict with tourism, the ecosystem and other coastal economic activities. Research and innovation in aquaculture, is also key.

The aquaculture industry is a very good example of commercialized upscale primary production. It has been a successful example where outside (of primary production) finances found their way into an industry which organized in a professional fashion, and utilized the comparative advantage of the region. This is an example that must be nourished and lessons have to be drawn to other industries as well. The greenhouse horticulture could be one such example, as well as others in livestock, dairy, poultry and others. Aquaculture does not need direct assistance, rather a favourable business environment, and consistent well-planned, transparent, integrated environmental rules. It also needs research support, which must be developed with thoughtful private–public partnership and link to the university research network.

The country needs such extraverted business, along with traditional and small-scale primary production. One should not exclude the other, and one can learn from each other. In particular, primary production can learn from the business scheme, financial innovations and management of the aquaculture.

## 5.4   Olive Oil

Olive oil constitutes more than 1.4% of total Greek exports. Greece is the third largest olive oil producer in the world, with an annual production of 385,000 tonnes (12-year average). Spain and Italy are the largest world producers with 1.1 million and 665 thousand tonnes average annual production, respectively. Greece exports an average of 14,000 tonnes annually, whereas the world's largest exporters Italy and Spain export annually 184,000 tonnes and 111,000 tonnes of olive oil, respectively. Greek olive oil production and exports are characterized by high cost of production, poor marketing infrastructure and reliance on bulk market (USITC 2013) (Fig. 5.15).

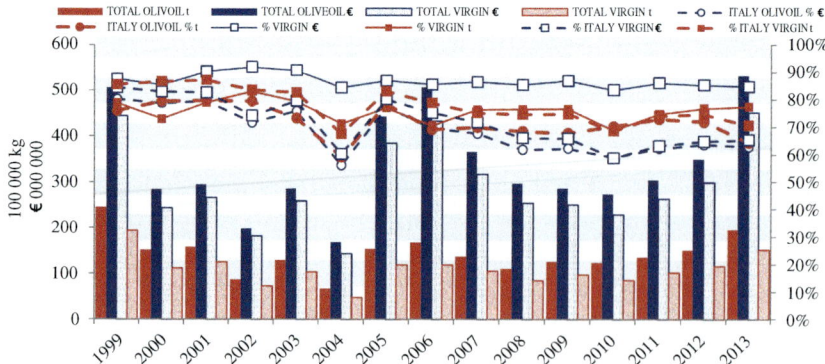

**Fig. 5.15** Exports of olive oil, Greece: 1999–2013. *Source* EUROSTAT, own calculation

Ironically, the main importer of Greek olive oil is Italy: more than 70% of quantity and more than 60% of value of total exports of olive oil (Fig. 5.16). Second largest importer of Greek olive oil is Spain (about 10% of value). Greece exports more than 80% of its olive oil to the two world's largest olive oil exporters, Italy and Spain.

Italian exporters blend, package and re-export the Greek olive oil (Fig. 5.16). Greece exports olive oil in bulk, mainly to Italy and Spain. A small portion of Greek olive oil finds its way to other export destinations of higher value. Italian exporters blend Italian olive oil, with Greek and Spanish and re-export. Ironically, a portion of the Italian blended olive oil is re-exported back to Greece (3.8%). Spain also is a major provider of Italian olive oil exports, since 24.4% of its total exports

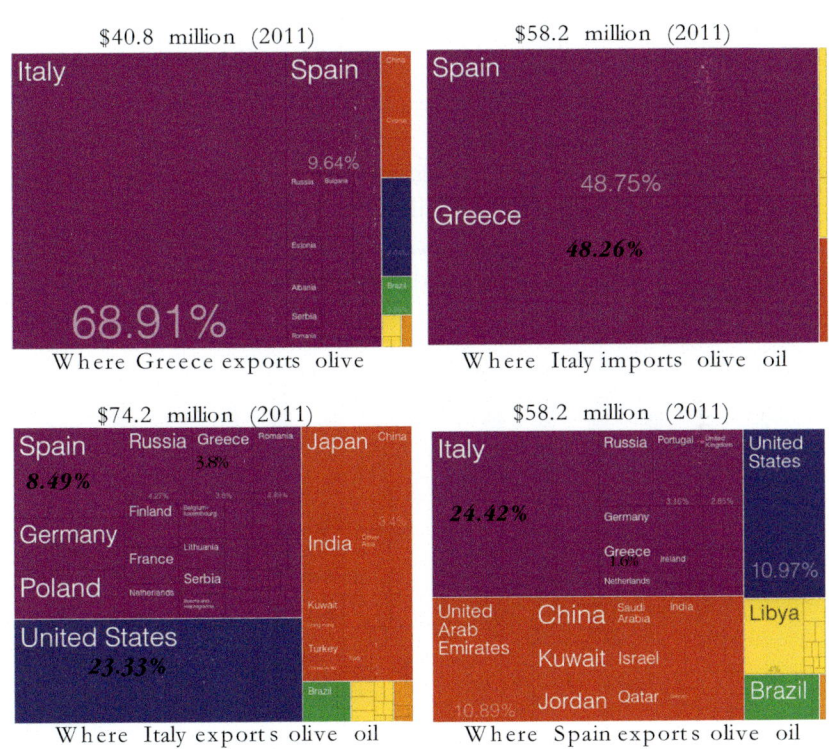

**Fig. 5.16** The trail of Greek olive oil exports (2011). *Source* http://atlas.media.mit. edu/explore/tree_map/—own construction

are destined to Italy. Spain, however, exports to other destinations such as USA, UAE, China and others. Also, a small portion of Spanish olive oil exports finds its way to Greece (1.6%).

Greek processing is taking place mostly in a large number of small scale, mills. There exist approximately 2200 olive oil mills in operation (EC 2012, p. 5). Compared to 5000 mills in Italy and 1740 mills in Spain, the average processing of a Greek olive oil mill is 295 tonnes, whereas that of Italy is 90 tonnes and Spain is 927.2 tonnes (based on 2011/12 production figures).

About 60% of the olive oil producers in Greece belong to a cooperative, while cooperatives own 20% of the mills (USITC 2013, p. 6–37). Yet, cooperatives suffer from low prices and difficulty to match supply with demand. Moreover, while they handle most of local production, they have not been able to gain significantly from processing and branding activities (Iliopoulos et al. 2012). Greek export prices are consistently below those in Italy (Fig. 5.17). Consequently, income of olive oil farmers in Greece is below national and EU levels (Fig. 5.18).

Greek olive oil is processed immediately after harvest, which raises the quality of olive oil. Many of the small mills have traditional sales channels and sell directly to consumers. Because of favourable climate and soil conditions, olive tree varieties and short time between harvest and

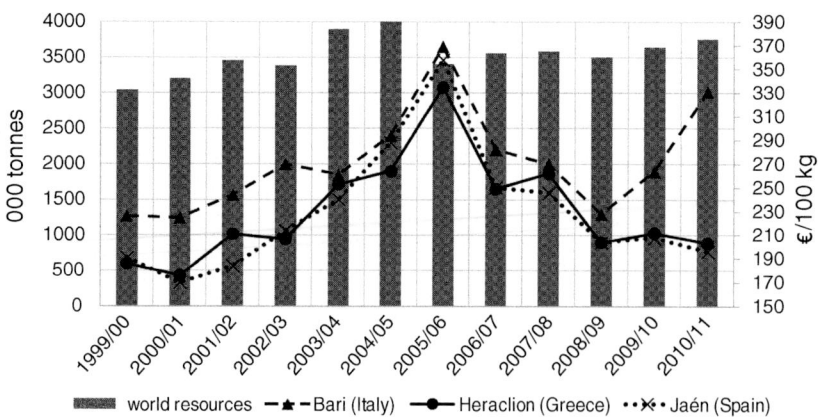

**Fig. 5.17** Prices of Olive Oil: Greece, Italy and Spain (1999–2011). *Source* Olivæ (2012)

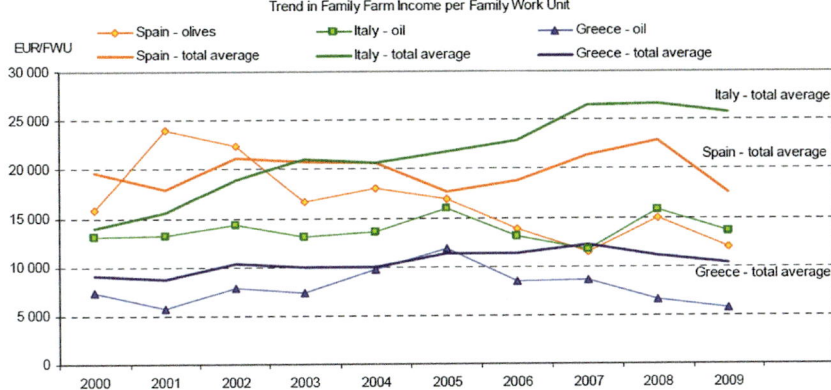

**Fig. 5.18** Income per family of olive oil farms. *Source* EC–DG AGRI (2012). Economic analysis of the olive sector

milling, Greek olive oil is considered of top quality by bottlers and distributors (USITC 2013, p. 6–37).

Despite its highly acclaimed quality, bottled Greek olive oil has not yet been developed to reach the international acclaim that many expect it should have. For example, in the recent 2014 international competition for olive oil in New York, Greek olive oils had a success rate of only 14.8%, compared to 50.3% and 46.8% for Spain and Italy, respectively (http://www.bestoliveoils.com/explore). Greece was the third largest competitor with 128 entries, but came fifth in the total ranking and did not manage to receive any "Best" award in any category.

## 5.4.1  Local Procurement—Mediterranean Diet—Tourism

With an initiative that started with the Heraklion Chamber of Commerce and Industry, the four Cretan Chambers of Commerce promoted voluntary agreements between local producers and the islands' hotels to use local products. The role of the Chambers of Commerce and Industry, as facilitators of such initiatives, is key. The lesson learned is that there are synergies in the collaboration between the three sectors:

primary (agricultural production), secondary (food processing) and tertiary (tourism, logistics).

In many respects, the olive oil industry follows closely the path and methods of the wine, and if the wine industry is the mode, it will not be too late until "new countries", such as Australia, South Africa, USA, Chile and others, outperform the traditional Mediterranean leaders (The Economist 2012).

The Mediterranean diet of Spain, Greece, Italy and Morocco was included in UNESCO's list of Intangible Cultural Heritage of Humanity in November 2010. Studies show that tourists prefer local diets (World Tourism Organization 2012; Velissariou and Vasilakaki 2010). As Greece tries to increase its exports of olive oil, Greece should also exploit its tourism and the Mediterranean diet, in order to increase the consumption of olive oil by foreign visitors and potential olive oil consumers, while they are in Greece.

Greeks are the world's largest consumers of olive oil, with 18 kg per capita annual consumption—or 50 g per person per day (Olive Oil Times 2013). The 2014 was a record year for Greek tourism with more than 22 million arrivals. If we estimate an average of 20 million tourists, with an average stay of 10 nights, this makes 200 million person-nights. If these tourists cater into the Greek Mediterranean diet, they should consume 10,000 tonnes of olive oil each year. This is close to the total of Greek olive oil exports. This does not add the usual trade costs of transportation, logistics, intermediaries, etc.

The Greek government announced recently a grant plan to increase tourist arrivals to 40 million by 2025 (TO BHMA, 3 May 2014). Even if this number is unrealistic, a doubling of the number of tourists is not entirely out of reach. If we assume that 50% of the total tourist days eat the Mediterranean diet, it will mean a doubling of olive oil exports. This does not count for the purchases of olive oil that tourists will take back home, neither for the follow-up sales that tourists will continue when they are back home.

We use this example, perhaps in exaggeration, in order to illustrate the importance of the synergies between tourism and the agri-food sector. It is perhaps not feasible to achieve such sales of Greek olive oil via the country's tourist sector to the levels illustrated here. It is based on the assumption that all tourists switch to Mediterranean diet and all restaurants and hotels use Greek olive oil. The example is used to emphasize the need of a national plan and coordination of the two sectors, tourism and agriculture. First, quality has to be improved, certified and be made consistent. Secondly, hotels and other tourist business, such as bed-and-breakfast places, restaurants, taverns, cruise ships, etc., need to adopt and develop further the Mediterranean cuisine. These tourist outlets need to further concur into consuming Greek food products, such as olive oil, fruits and vegetables, dairy, meat and legumes. What is even more important is the enhancement of post-visit sales. Tourists can buy and bring olive oil—and other food products—back home. Most importantly, they should be able to find olive oil and the other products in their local market when back home. This requires coordination, branding and logistics; it requires a concerted action, which has very high costs and large economies of scale. This activity is best undertaken by a central organization, such as the Greek Inter-Professional Agriculture and Food Council (GIPAFC) and the Greek Logistics Hub (GFLH)—both presented in Sect. 6.2 and Fig. 6.4.

The olive production and olive oil extraction are at the heart of Greek agriculture, culture and history. These are traits that must be nurtured and promoted to their highest value. Due to these very specialized features of the terrain, microclimate, history and culture, olive oil and table olive production are varied and exhibit local idiosyncratic characteristics that may fetch high benefits to producers, the local economy and the national economy as a whole. These idiosyncrasies should not be collapsed and disappear into bulk generic commodity production, since they will lose their intrinsic value. Instead of promoting large-scale processing facilities, a Greek olive oil industrial strategy should encourage product differentiation, based on organoleptic and other attributes specific to each case. Of special importance is to link olive oil production, processing and promotion to the promotion of Mediterranean diet and the tourist industry, where this is feasible. To accomplish this, large scale at the primary level is not necessary. Neither processing needs to be at large industrial scale. Instead, local, often traditional, techniques should be elevated to higher technological standards, but in ways that retain the high quality and the specialized characteristics of each variety, terrain, history and heritage, in strategic ways that add value and high returns.

One (or maybe two) large facility countrywide to process the country's olive oil production as suggested recently (McKinsey & Company 2012) is an extreme proposition. Greece's olive oil production has a wide dispersion in geography and variety and any paradigm of the olive oil sector should account for this. This is true for most of the Greek agri-food production, and both large- and small-scale facilities should coexist. There is room for both large and smaller facilities, and competition in scale, quality and differentiation should determine the optimum size and distribution. Large-scale facilities may provide the avenue for the lower quality product and will enhance the purity and promotion of higher quality which can be sold at higher prices.

## 5.4.2 Olive Oil, Wine and Agrotourism: Complexity and Isomorphism

The new paradigm needs to exploit the potentials of complexity and isomorphism that are inherent in most, and especially the Greek agri-food system. Complexity, is measured by the diversity of capabilities present in a country and their interactions (Hidalgo and Hausman 2009). The complexity of an economy, or a sector, is related to the multiplicity of useful knowledge embedded in it. Interaction between individuals through complex networks is necessary in order to make competitive products and develop competitive industries. Isomorphism is the tendency of firms to follow similar processes and organizational forms (DiMaggio and Powell 2000). We argue that both forces need to be integrated into the new paradigm for Greek agriculture. Knowledge and capabilities vary between industries but can be integrated and adapted by each other. Successful processes and business models can be adopted as well. We use two cases from the olive oil and wine industries to exemplify the significance of complexity and isomorphism. Biolea is a family business producing virgin olive oil in Crete, and Gerovassiliou Estate Winery is a successful wine business in northern Greece.

These two cases exhibit the potential for cross-fertilization of ideas, and business models—a sort of isomorphism—and the integration with tourism. Both cases indicate that the existing institutional framework often acts as a constraint, rather than as an aid to growth and development of new business ideas.

Biolea is a family-owned, vertically integrated organic olive oil production unit in Astrikas, Crete. It was established on a 10-ha olive grove. The infrastructure was built with local architectural characteristics and the ability to facilitate visitors. A stone mill, upgraded with modern technology, and hydraulic cold press, with in-house bottling line produces high-quality olive oil sold at the high end of the price spectrum. All operations run under environmental ISO 14001. Biolea Estate offers guided tours and olive oil tastings to over 8000 visitors per season.

Gerovassiliou is a winery in Epanomi, Thessaloniki. All domaine wines are produced from grapes cultivated in the private vineyard, now stretching over 56 ha. It has received numerous international awards and gold metals. The Estate has a unique wine museum and offers organized visits, wine tastings, educational programmes, seminars and publications and has recently built a restaurant for the visitors and for hosting events. Mr. Gerovassiliou with two other winemakers and marketers has recently created a small winery, ESCAPADES WINERY, at Stellenbosch, South Africa.

"We should not go back to tradition—we should bring the tradition forward to today's terms". This was the emphatic motto of Mr. Dimitriadis, Biolea's founder. He also revealed that his model for the olive oil estate was inspired by the wineries he had visited in his travels around the world, especially in Italy and France. Taking advantage of the tradition and local (even the family history and heritage), the local comparative advantage, and combining it all with modern technology, high aesthetics and tourism, was the recipe of success.

Both businessmen's main complaint was on institutional constraints posed by government bureaucracy, anachronistic legal system and lack of vision. Biolea's traditional millstone and cold press oil refinery could not be built and licensed in the estate until an old deserted oil press was purchased simply to acquire the license—a very complex, time-consuming and expensive process. Similarly, Mr. Gerovassiliou is struggling to acquire a license for the restaurant they have recently built on the Estate. In addition, wineries have to struggle with multiple authorities, the GMRDF, the General Chemical State Laboratory and the Customs Authority.

It is evident from these cases that knowledge and capabilities from various industries need to be exchanged in order to help cross-fertilization and growth. The organizational structure under the new paradigm as proposed here is one way to allow synergies and minimize transaction costs.

## 5.5    Organics and Local Products PDO and PGI

There are 55 certified protected designation of origin (PDO) products in Greece and 23 product with geographic indication (PGI) products (Table 5.1). Sales for PDO and PGI products are increasing. Production and sales of organic products (Fig. 5.19) and PDO and PGI products (Figs. 5.20 and 5.21) have increased in Greece during the last decade.

**Table 5.1** Protected designation of origin (PDO), geographic indication (PGI): Greece 2014

|                      | PDO      | PGI       |
| -------------------- | -------- | --------- |
| Cheeses              | 21       |           |
| Fruits and vegetables| 9        | 12        |
| Fruits               | 7        | 4         |
| Vegetables           | 1        | 2         |
| Pulses               | 1        | 6         |
| Olive oils           | 17       | 11        |
| Animal products      | 3        |           |
| Essential oils       | 1        |           |
| Fish products        | 1        |           |
| Natural gums         | 2        |           |
| Bakery products      | 1        |           |
| Total                | 55       | 23        |
| *Wines*              | *29 zones* | *103 zones* |
| Regional             |          | 8 zones   |
| District             |          | 37 zones  |
| Area                 |          | 58 zones  |

*Source* GMRDF, http://www.newwinesofgreece.com —own calculation

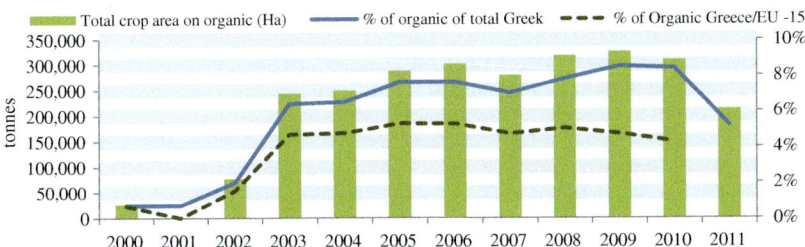

**Fig. 5.19** Area under organics: Greece 2000–2011. *Source* EUROSTAT, own calculation

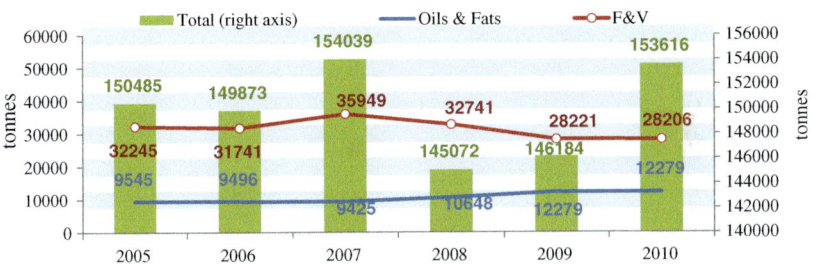

**Fig. 5.20** Sales volume PGI and PDO: 2005–2010. *PGI* Product with geographic indication, *PDO* Protected Designation of Origin. *Source* DG-AGRI, own calculation

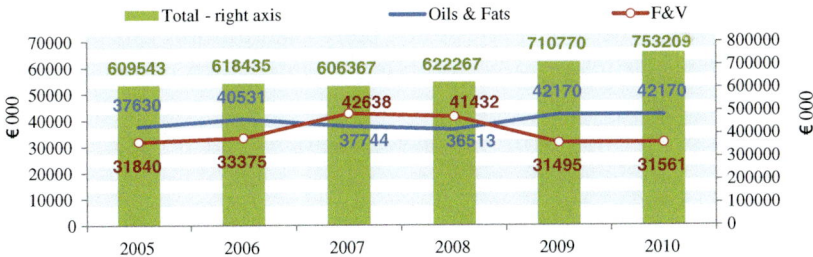

**Fig. 5.21** Sales value €PGI and PDO: Greece 2005–2010. *PGI* Product with geographic indication, *PDO* Protected Designation of Origin. *Source* DG-AGRI, own calculation

Organic farming in Greece began before any regulatory framework existed (Greenfood 2010). The first organic production in Greece began in 1982 in Aigialeia where a small group of local farmers started organic Corinthian grapes for export to Holland. No more than 200 ha on organic were grown until the 1990. EU regulations 2092/91 and 2078/92 and 1804/99 brought in annual growth rates of up to 120% in organic farming. Today, organic production is about 0.5% of UAA and 0.5% of the total number of holdings (24,000 farms in 2009) (Greenfood 2010). Most of the organic area is grazing lands (47%), fodder (14%), olives (18%) and cereals (12%). Fruits and vegetables account for about 5%. Organic production shows indications of decline after 2010 (Fig. 5.19). The growth and subsequent decline are attributed mainly to subsidies (EC 2014b, p. 13).

Greece could expand organic production much further. Much of olive production, fruits and livestock, are produced with traditional near-organic and environmentally friendly methods for centuries. Expansion, however, is limited by lack of organization, information and training. Still, there is a tremendous potential for expansion.

A total of 55 PDO and 23 PGI products are recognized in Greece, as well as 29 and 103 wine zones, respectively (Table 5.1). Feta cheese is prime example of PDO. After numerous legal battles, the Greek Feta has received the PDO label and is prime export item. The full potential of Feta has not been exploited; however, since competition among exporters is threatening to turn this product into a commodity, a national strategy is needed here. The inter-professional dairy organization can play a key role, and the stakeholders need to understand that such national strategy is to their advantage.

Notable is the success of Greek yogurt. Based on private label, one family-owned company succeeded to establish a global brand name. The "Greek yogurt" brand name is the creation of FAGE. This is a success of three generations of entrepreneurs of the family behind FAGE. The first yogurt shop was established in 1926, in Athens. Production expanded and the first nationwide yogurt supply chain was established by the family-owned company in the 1950s. This led to a factory in Galatsi—a suburb of Athens in the 1960s. Sales grew fast, and in the 1970s, a new larger facility was established. The FAGE yogurt was shipped to smart packaging throughout the country. Exports of FAGE yogurt started in the 1983, first to the UK, then to Italy and to the USA in 1998. Today, FAGE exports its yogurt to 33 countries. In 2008, FAGE established a yogurt production facility in Johnstown, New York. The Johnstown facility is the largest among the four facilities (3 in Greece) owned by FAGE today. Other firms have been using the brand name "Greek yogurt", some of them with great success (e.g. the CHOBANI brand in the USA, who is the primary contributor to Greek yogurt's rise in the USA).

The Greek yogurt and Feta are two cases that a lot can be learned from both of them. The yogurt, was developed by a private firm, almost single-handedly—not a collective action. The establishment of Greek Feta, on the other hand, as a PDO product was the result of national effort, primarily by the GMFRD and by private interests. The Feta is produced by many producers of various scales of production. In order to maintain and increase the growth of Feta and other Greek cheeses (some of them, observably of superb quality), concerted action is needed. The dairy inter-professional organization, the GOMM, and especially the GIPAFC can play a key role for the further development of Feta.

## 5.6  Summary and Recommendations

### 5.6.1  Aquaculture

The aquaculture industry needs to work close with banks for the development of instruments to help overcome the current credit bottleneck and avoid similar problems in future. Aquaculture firms need to consolidate to efficient scale, where appropriate, in order to reduce costs of production. Aquaculture inter-professional organization (AQIPO) should be formed, to become part of the GIPAFC which will provide services and strategy. The GMRDF in consultation with the AQIPO, and related organizations, regions, municipalities and the tourist sector, must very quickly promote legislation that resolves the legal status, zoning and application procedures for the development of aquaculture. Development of promotion of brand name for Greek fish by the AQIPO and in collaboration with GIPAFC is essential. A proper and effective franchise certification of Greek aquaculture products and Aquaculture Research and Development Fund (AQRDF) from a small fee paid per unit of fish sold should be created.

## 5.6.2  Olive Oil

The Olive Oil industry should coordinate and synchronise strategy in close collaboration with tourism, and restaurants and be an integral part of the national plan for the promotion of the Mediterranean diet. Olive inter-professional organization (OLIPO) should be formed, to become part of the GIPAFC which will provide services and strategy. The local production with specific organoleptic characteristics, terrain, history and cultural heritage must be promoted. Promotion of brand name for Greek olive oil by the OLIPO in collaboration with GIPAFC must be developed. A proper and effective franchise certification of Greek olive oil and Olive Research and Development Fund (OLRDF) from a small fee paid per unit of olive products sold should be created. National Olive Research Centre (NORC) must be established to promote research and development for primary production, breeding, plant protection and agronomic practices, as well as the development of improvement of processing and promotion of olive products. The larger scale processing facilities at strategic areas, such as Peloponnese, Crete, Attica and perhaps elsewhere, will process and promote quantities that are not able, or are not desired to be processed at the local specialized level. Educational and promotional and awards programme for culinary professional on olive oil appreciation—similar to wine appreciation—must be conducted.

# References

DiMaggio, P.J., and W.W. Powell. 2000. *The iron cage revisited institutional isomorphism and collective rationality in organizational fields*, 143–166. Economics Meets Sociology in Strategic Management: Emerald Group Publishing Limited.

EC. 2014a. Report from the Commission to the European Parliament and the Council on the Implementation of the Provisions Concerning Producer Organisations, Operational Funds and Operational Programmes in the Fruit and Vegetables Sector since the 2007 Reform. EC COM (2014) 112 final.

EC. 2014b. Commission Staff Working Document. Impact Assessment Accompanying the Document Proposal for a Regulation of the European Parliament and of the Council on Organic Production and Labelling of Organic Products, Amending Regulation (EU) No. XXX/XXX of the European Parliament and of the Council [Official controls Regulation] and Repealing Council Regulation (EC) No 834/2007.

EC–DG AGRI. 2012. Rural Development in the EU Statistical and Economic Information Report.

Economist. 2012. Olive-oil proces – Drizzle and drought. September 22.

ELSTAT. 2014. Hellenic Statistic Authority. Various years. http://www.statistics.gr/.

Greenfood Project. 2010. Analysis of Organic Agrarian Activity in Greece.

Hidalgo, C.A., and R. Hausmann. 2009. The Building Blocks of Economic Complexity. *Proceedings of the National Academy of Sciences* 106 (26): 10570–10575.

Iliopoulos, C., C. Giagnocavo, I. Theodorakopoulou, and S. Gerez. 2012. *Structure and Strategy of Olive Oil Cooperatives: Comparing Crete*. Greece to Andalusia, Spain: Support for Farmers' Cooperatives.

Jordi, G., and M. Motta. 2013. The Economic Performance of the EU Aquaculture Sector (STECF 13-29)-Scientific, Technical and Economic Committee for Fisheries (STECF). EU.

McKinsey & Company. 2012. Greece 10 Years Ahead. Athens.

Netherlands embassy in Athens Greece. 2012. Developments in the Greek horticultural sector.

Olivæ. 2012. International Olive Council. v. 117.

Olive Oil Times. 2013. Greeks Still World's Top Olive Oil Guzzlers. June 6.

Reuters. 2013. Insight: In Greece, fish farms a testing ground for economic revival. http://www.reuters.com/article/us-greece-fishfarming-insight-idUSBRE 9BF09920131216.

USITC. 2013. Olive Oil: Conditions of Competition between U.S. and Major Foreign Supplier Industries. USITC Publication 4419.

Velissariou E., E. Vasilakaki 2010. Local Gastronomy and Tourist Behavior: Research on Domestic Tourism in Greece. 4th International Conference on Tourism & Hospitality Management, Athens, Greece.

Vlachos, G. and Karanikolas, P. 2013. Adjustment for Survival: The Case of Peach Producers in Imathia (in Greek) International and European policy 27, 50–59.

World Tourism Organization. 2012. *Global Report on Food Tourism*. Madrid: UNWTO.

# 6

# Framing a New Paradigm for Greek Agriculture

**Abstract** The new paradigm for Greek agriculture requires a pyramid organization that will implement the necessary changes in the Greek agri-food system. The basis of the new system is the Greek Inter-Professional Agriculture and Food Council (GIPAFC). The GIPAFC is founded on four pillars: inter-professional organizations, cooperatives, research and training, related sectors. It incorporates committees and integrates into organizations that handle among others, an export hub for Greek food products.

**Keywords** Paradigm for Greek agriculture · Inter-Professional organizations · Governance · Experience economy

© The Author(s) 2017
K. Karantininis, *A New Paradigm for Greek Agriculture*,
DOI 10.1007/978-3-319-59075-2_6

## 6.1 Framework of Analysis of Greek Agri-Food Value Chain

Many of the problems and drawbacks in the growth of Greek agriculture and food are of institutional nature. Their solution demands actions that are not necessarily costly financially—perhaps more costly on the political side, since decisions and break-ups with past headlocks are necessary.

To frame the analysis, we follow a matrix value chain approach (Fig. 6.1): The agri-food chain is divided vertically into 12 levels, and each level is analysed in four dimensions.

**Vertically**, the value chain is divided into five upstream and six downstream levels:

| | size | | ownership | | governance | | | | space | | | |
|---|---|---|---|---|---|---|---|---|---|---|---|---|
| | small | large | private | public | atomistic | hierarchical | collective | contractual | local | regional | national | global |
| policy & lobbying | | X | | X | | | X | | | X | X | X |
| research | | X | X | X | | X | | X | | X | X | X |
| education & training | | X | | X | | X | X | | X | X | X | X |
| extension | | X | | X | | X | X | X | X | | X | |
| farm inputs | | X | X | | | | X | X | X | | X | X |
| FARMING | X | X | X | | X | X | | X | X | | | |
| wholesale | | X | X | | X | X | X | X | | X | X | X |
| processing | X | X | X | | X | X | X | | X | | X | |
| distribution | | X | X | | X | X | X | X | | | X | |
| exports | | X | X | | | X | X | | | | | X |
| promotion | | X | X | | | X | X | | | | X | X |
| retail | X | X | X | | X | X | X | | X | X | X | |

Fig. 6.1 The value chain and dimensions of organization. *Source* Own construction

*A. Upstream levels*:

- Policy and lobbying
- Research
- Education and training
- Extension
- Farm inputs.

*B. Downstream levels*

- Wholesale
- Processing
- Distribution
- Exports
- Promotion
- Retail.

**Horizontally**, each level is analysed in four dimensions, and each dimension is divided into two or more discrete levels:

- **Size**: Small–large
- **Ownership**: Private–state
- **Governance**: Atomistic–hierarchical–collective–contractual
- **Space**: Local–regional–global.

**Farming**: In terms of size, farms are small or large. The ownership of farms is private in their majority, but public lands also exist, mainly in forestry and pastures. Farm governance is mainly atomistic and market-oriented (family farm); however, contractual production is increasing, while we have very few collective (cooperative) farms. Spatially, farms are mostly local; however, we have some farms that extend to various regions in the country, and recently, we even have farmers that cross borders becoming global.

**Policy and lobbying** needs to be large, public, collective and at national or global (EU) levels.

**Research** is best accomplished at large scale, in combination with private and public ownership and collective and/or contractual governance, at a national level.

**Education and training** needs large scale; it can be both private and public, individual or collective, mostly at national level.

**Extension** is currently both small (individual agronomists) and large (national extension service) and both private (individual agronomists) and public; it is also hierarchical as well as atomistic, at local, regional and national levels. Extension would function better if it was at large scale, public and collective at national level.

**Farm inputs** (chemicals, fertilizer, credit, etc.) are currently large, private, atomistic, and national or global (multinationals).

**Wholesale and processing** are small and large, private, atomistic or collective, at local, regional or national, and global levels.

**Exports** are large; private; atomistic, hierarchical or collective; and global.

**Distribution, promotion and retail** are mostly large, atomistic or hierarchical, and some are collective, local, regional, national and global.

Any sustainable agri-food value chain needs, primarily, to take into account the special characteristics of the demand for its products. The demand for food is increasingly driven by the desire of consumers to acquire experience attributes from the goods they consume. The so-called new experience economy is discussed next in Sect. 6.1.2.

## 6.1.1 The Experience Economy

Modern affluent economies have evolved from agrarian economies to industrial economies in the nineteenth century to service economies in the late twentieth century. Consumers in affluent societies expect more than simply quantity and quality in the consumption of goods they consume, whether these are cars, mobile phones or beverages and food. Consumers expect further experiences. The evolution to an "experience economy" is shown by the growth of consumption of leisure activities, entertainment, travel and tourism, while the share of consumption of commodities is declining (Pine and Gilmore 1999; Swinnen et al. 2012).

It has been argued that the future growth of European agriculture (Swinnen et al. 2012; Matthews 2012), and Greek agriculture in particular (Damianos and Vlachos 2014), lies not on efficiency and productivity alone, and on production of commodities, neither—especially —on further subsidization of primary agriculture, but on the "experience economy". A growing number of consumers are willing to pay a premium for food products that add some form of intangible experience. This experience is not exhausted with the consumption of the good, but continues beyond its consumption. Food products, such as organic, or "fair trade", as well as products with PDO, organic products and fair trade, are in this category. Furthermore, products that are sold in local markets, or on local or theme trade fairs, add further experiences. The advantage of such products to the producer is especially related to product differentiation and the feasibility of small scale.

The new experience economy poses challenges on policy and organization of the agri-food chain. The question is whether more resources need to be directed towards rural development and innovation programmes, and programmes that can assist farmers and the agri-food chain to redirect itself towards the experience economy—rather than continue support on commodity-oriented direct payments. Despite scepticism as to what extent specific support policies can drive such change (e.g. Swinnen, et al. 2012, p. 43; Matthews 2012), the potential of the experience economy must not be overlooked by policy-makers and agri-food stakeholders.

Particularly important is the role of the private sector, and its reorganization towards the experience economy. The value chain needs to reorganize and develop where necessary short supply chains; local markets must be developed, and branding of local and "experience products" must be put in place. The re-organization of the value chain is very key for two reasons. First, it is necessary to organize in order to supply the multitude of these products to final consumers, and transfer back to producers the information of the consumers' demands. Secondly, the chain must be in a position to transfer the premiums that will be generated efficiently to the producers and other stakeholders in a manner that is both efficient and equitable.

It would be naïve to argue that the Greek agri-food system should orientate entirely towards the production and distribution of experience goods. This is neither feasible nor desirable. Instead, the agri-food value chain needs to scale up reorientation towards product differentiation. A hybrid system that accommodates both large and small is desired, while the monolithic "race to the bottom" with only large-scale production must be avoided.

## 6.1.2 Product Differentiation, Scale and Scope: Need to Scale up

One comparative advantage of the Greek agri-food production is product differentiation and scope. The geography, microclimates, history and cultural diversity of Greece have resulted in a multitude of products, with local geographic and cultural characteristics. The success story of Greek yogurt, and the unrealized potential of feta cheese, are only of many examples. The downside of this comparative advantage is the small size which is associated with the following:

a. High production costs at primary level, due to lack of scale economies
b. High transaction costs, of coordination, and enforcement of contracts
c. Lack of consistency and continuation
d. Low innovation capacity—mainly due to scale and lack of national strategy.

a. *High production costs at primary level*

In general, small scale is associated with higher costs of production, but not necessarily with inefficiency. There is no clear scientific evidence pointing towards higher efficiency of large farms (see, e.g., Kaditi and Nitsi 2010). Furthermore, small scale is very often inevitable, especially in mountain and island regions. The challenge, therefore, is not so much how to increase further efficiency by larger scale—it is rather how to turn small scale into a competitive advantage. Even further, small and large scale need to coexist into a hybrid mode of production, processing and distribution of food.

While small scale is inevitable in mountainous and island areas, larger scale production in the plain areas of Macedonia Thrace and Thessaly is the mode of production in grains, cotton, etc.

It cannot be emphasized enough how important it is to scale up, i.e. to integrate both forwards and backwards and attain economies of scale in processing, distribution and logistics, research and marketing. The GIPAFC structure can be an instrument towards scale-up.

### b. *High transaction costs*

Transaction costs are the costs of doing business. Transaction costs are associated with uncertainty, bounded rationality, opportunistic behaviour and highly specific assets. The overall institutional environment of an economy adds to transaction costs by increasing further the costs of doing business. Greece ranks very low in terms of the efficacy of its institutions (Schwab 2013). The factors that contribute to high transactions costs are among others, bureaucracy, tax regulations, instability of policies and corruption (Schwab 2013).

Coordination of a scattered supply chain is cumbersome and involves high transaction costs. Coordination costs can be lowered with scale-up and standardization and dissemination of information. Vertical structures such as those under the GIPAFC can move towards this direction.

### c. *Inconsistency and discontinuity*

Many of the local products are often not commercialized at all, or if they are, it is done in a small, local scale, which is not consistent, standardized or branded. High uncertainty and low entry costs (these products often do not require high investments) lead small producers to enter and exit. This variability makes long-term planning along the chain difficult and costly.

There is a large variety of small-scale production of artisanal products, such as marmalades, sweets, salads, bakery, cheeses and sausages, which due to the very small scale and inconsistency in both quantity and quality, are not interesting for larger distribution channels. These products, however, constitute a potential to support rural development. There is need to scale up. This can be done with these very small production units to integrate via producers groups and cooperatives and IOFs of various sizes. Standardization, branding and access to local and export markets through a logistics centre, such as the GFLH (*Greek Food*

*Logistics Hub*) under the GIPAFC, are crucial steps. Similarly, access to organized farmers markets organized under the CDFM (*Centre for Development of Farmers' Markets*) will provide an avenue for the development of the differentiated products.

d. *Low innovation capacity*

Small firms do not have the capacity to innovate at the same rate as their larger competitors. They need scale-up. Continuous training, highly efficient and flexible extension service access to finance and access to information are necessary. Synergies with other firms, not only food-related firms, are key. The innovation parks (FIP) and firm incubators (AFFI) under the GIPAFC can be venues of innovation, for small entrepreneurs as well as larger firms.

## 6.1.3  Governance Structures

We briefly discuss here some basic governance structures, such as markets, hierarchies, collective structures (cooperatives) and contracts.

a. *Atomistic markets, futures markets and farmers' markets*

    i. Atomistic markets

       The atomistic (spot) market is the fundamental institution of the capitalist system. The role of markets in coordination, price determination and price discovery cannot be emphasized enough. The state must provide the appropriate regulatory framework for the good functioning of all markets in the agri-food chain. The state must also monitor the competitive behaviour of firms with proper instrumentation, such as the *Competition Committee* (*ΕπιτροπήΑνταγωνισμού*) and other ministerial agencies.

    ii. Futures markets

       The use of futures markets is not very much spread among the Greek agri-food industry. With the exception of some large— mostly multinational—firms, futures commodities markets are unknown. This is mainly due to ignorance and also size. Futures and other commodity derivative contracts need both good knowledge of the particulars of markets and also a certain size. These markets could be useful to traders and processors in commodities such as grains and cotton. Individual cotton farmers, for instance, could use hedging strategies with the proper use of commodity futures. Here, the role of collective organizations, such as cooperatives, cooperative unions and the inter-professional organizations (IPOs), could play a key role. This potential is unexploited in Greek agri-food. The potential of futures markets can be further exploited through research, and training, and later by creating appropriate entities that will involve in futures trading on behalf of interested farmers and other agri-food stakeholders.

iii. Farmers' markets

Farmer's markets must be promoted and safeguarded. These markets can operate in collaboration with local and municipal authorities and should be extended not only in fresh produce but also processed products, dairy, fish and meat. The GMRDF should establish rules and authorities to guarantee food safety and quality of the products that are distributed through these markets. The private agri-food sector, through the GIPAFC, must promote, develop and safeguard farmers' markets, through organizations such as the the Centre for Development of Farmers' Markets (CDFM).

- One model of a farmers' market organization can be a special-purpose cooperative (*Local Farmers' Market Cooperative —LFMC*) for each market where the local municipality can be a co-owner and be represented on the board (a private firm (IOF) with similar purpose and representation should not be excluded). This cooperative will be in charge of hiring space, collecting fees and monitoring the operation of their own farmers' market. Each farmer shall be allowed to be a member of more than one such cooperative, but can be an elected representative to only one. Members will be given a priority for space at each market. We envisage a number of such special-purpose cooperatives, which can establish a National Federation of Farmers' Markets' Cooperatives (NFFMC), and can be represented at the national umbrella inter-professional organization GIPAFC. This national federation of farmers' market cooperatives can create a monitoring and development body (such as the CDFM) with professionals who will monitor and help the promotion, training, marketing and merchandizing of farmers' products at these markets.

- Specialized farmers' markets, festivals and bazaars should be established in touristic areas during the tourist season. They should be designed to cater the tourist clients and strive to create loyal customers when tourists return to their homes. The NFFMC through the CDFM can help to establish such special tourist farmers' markets and help in the follow-up, in close collaboration with the GPAFC and the GFLH.
- Specialty marked farmers' stalls can also be established at the local open markets.
- Production and processing of artisanal scale must be promoted. The quality and food safety of the products must be monitored and guaranteed by the local authorities and overseen by the GMRDF, the CDFM and the GIPAFC.

b. *Collective governance: cooperatives and producer groups*

More detailed discussions on Greek cooperatives are in Sects. 4.6, and 5.2.4. In general, the cooperative organizations need to rid themselves from the sins of the past; they need to reorganize, restructure and modernize with professional management. They should open to outside investors and form alliances with other coops and IOFs.

i. **Producer groups (PG) (Ομάδες Παραγωγών)**
   Producer groups (PG) (Ομάδες Παραγωγών) are now covered under article 5/4015/2011. This law allows the formation of PGs and restricts direct sales by farmer members to consumers up to 10% of their production.

ii. **Investor-owned firms (IOFs) and other hierarchies**
Farmers should not be restricted to the cooperative form—although it is a preferred form of such business—but should also be allowed to form private or limited liability companies to process and trade their products at small and larger scale. For this purpose, the establishment of such firms must be simplified and become more inexpensive. The laws of competition and maintenance of quality and food safety must apply.

iii. **Inter-professional organization (IPO) and inter-branch organization (IBO)**
The IPOs are key to the success of the future of Greek agriculture and food. They pass responsibility to stakeholders, who organize, and coordinate, research, promotion, monitoring and extension. IPOs can be established according to recent legislation (article 8/4015/2011). The 8/4015/2011 allows only one IPO at the national level for a product or group of products.
What is missing from this legislation is the rules for the formation of an umbrella IPO (**Greek Inter-Professional Agri-Food Council—GIPAFC**) which will cover all IPOs, cooperatives and other stakeholders of the sector. In order to develop a sustainable agri-food value chain, an adaptive organizational structure is necessary to construct a strategy and implement the necessary actions. This organization needs to be representative and have the appropriate scale and scope. The **GIPAFC**, presented below, has these virtues. It needs immediately the direct involvement of stakeholders and the appropriate legislative framework.

## 6.2 The Greek Inter-professional Agri-Food Council (GIPAFC)

The GIPAFC is a pyramid organization (Fig. 6.2). Its membership will consist of all involved stakeholders of the agriculture and food industry in Greece. The key roles of GIPAFC are as follows:

- Coordination along and across the agri-food value chain
- Consensus building within the sector
- Political representation of the sector's collective interests
- Business representation of the Greek agricultural sector nationally and internationally
- Provision of services to its members:

  - Research
  - Education and training
  - Extension
  - Business and legal counselling and conflict resolution
  - Development and safeguarding of a Greek food brand name
  - Promotion of Greek food products abroad via a logistics hub
  - Assist and promote innovation in the agri-food sector

- Link the sectors' interests with other sectors, such as tourism and other industry

### 6.2.1 GIPAFC Pyramid Structure

The representation in the GIPAFC will be built on four pillars, which will represent the key groups of stakeholders (Fig. 6.2).

   I  Inter-professional organizations and inter-branch organizations: These will be the organizations formed according to article 8/4015/2011.

  II  Cooperatives and farms organizations: Cooperatives will be represented through the federal body (PASEGES or other body) as well as directly. Similarly, other farm organizations such as farmers' unions will be represented in II.

 III  Research, education, extension, training and extension: Here, there will be representation of universities, farm schools and research organizations.

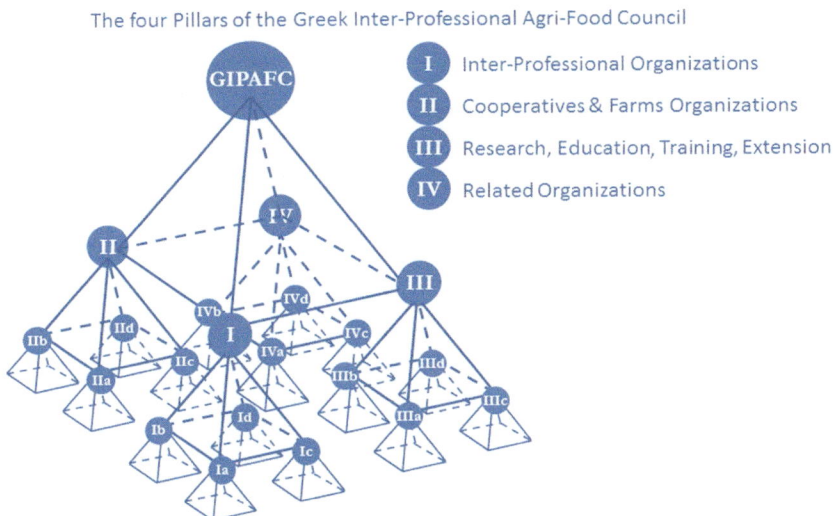

**Fig. 6.2**   The GIPAFC pyramid structure. *Source* Own construction

IV   Related organizations:

In order to promote synergies with other related sectors, there will be representation, for example, by tourism, restaurant organization, other industry, etc. Individual firms will be able to participate with approval by the board and by paying the fee. The exact structure, voting rights and fees of the GIPAFC are not elaborated here (Fig. 6.2).

## 6.2.2   GIPAFC Committees

The GIPAFC will be linked horizontally with the ministry (GMAFRD) and other organizations via committees (Fig. 6.3). Indicative committees will be

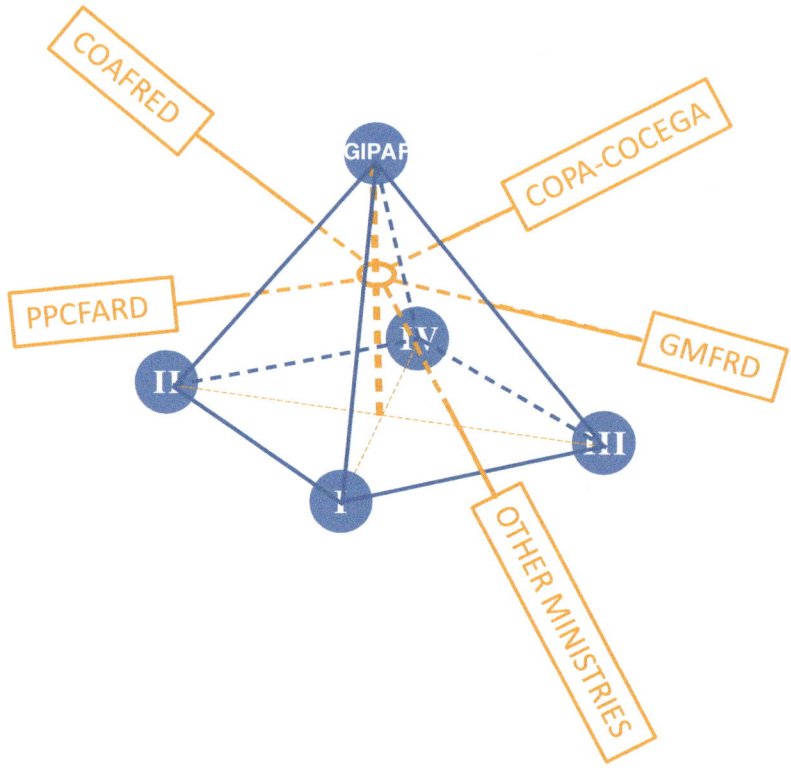

**Fig. 6.3** The GIPAFC committees. *Source* Own construction

a. A committee on policy with the ministry GMFRD, with representation by the PSAFR (Permanent Secretary of Agri-Food and Rural
Development)
b. Representation at the COPA-COCEGA
c. COAFRED—Committee on Agriculture Food Research and
Education
d. Committees with other ministries, such as environment, external
affairs, economy and trade.

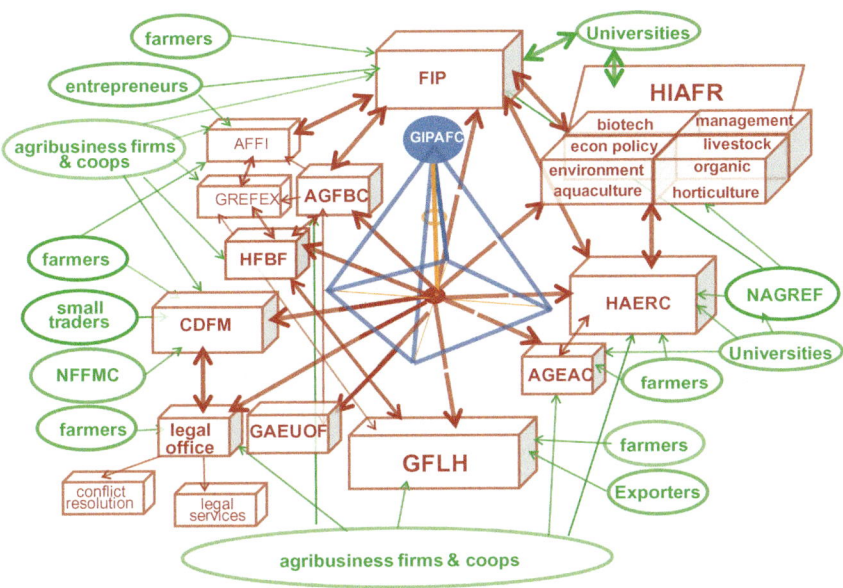

**Fig. 6.4** The GIPAFC structure. *Source* Own construction

## 6.2.3 GIPAFC Business and Other Entities

The GIPAFC will create various business entities and joint ventures (Fig. 6.4):

1. *HIAFR*—Hellenic Institute of Agriculture and Food Research: This is a very essential role that the GIPAFC must play in research in agriculture and food. The HIAFR will be financed by a fund that will be created for this purpose.
   The **HAFRDF**—Hellenic Agricultural and Food Research and Development Fund: The HIAFR will be linked, among others, with the universities, the NAGREF, the HAERC and the FIPs.
2. *HAERC*—Hellenic Agriculture Extension Research Centre: This is a highly needed centre for applied research in extension. It will support and educate the extension services around the country. The HAERC will support the AGEACs and will be closely linked with the HIAFR, the NAGREF, universities and directly with farmers.

3. *AGEAC*—Agriculture Extension and Advisory Centres: The GIPAFC must run its own high level, professional, model extension services. Extension needs to be upgraded and must be liberated from the hands of chemical and other similar companies and individuals who currently provide most of the private extension and advisory services, which are not objective, although often are served by highly skilled professional agronomists.

4. *FIP*—Food Innovation Park: The GIPAFC together with the GMAFRD and agribusiness firms will develop and facilitate thematic FIPs around the country. These will be places where innovation will stem out of synergies and agglomeration economies from collaboration with researchers and entrepreneurs.

5. *AGFBC*—Agriculture Food Business Centre: This centre will create and facilitate these very important entities.

   a. *AFFI*—Agri-food Firm Incubators: These will be linked to the FIPs and will provide young entrepreneurs with low-cost facilities and support to promote business ideas in the agri-food and start a new business.

   b. *GREFEX*—Greek Food Experience: The AGFBC is to promote the Greek food, the Greek and Mediterranean diet and to establish the commercial links with tourism. GREFEX will develop and support a Hellenic Food Portal (HFP) and the organization necessary to establish the HFBF.

   c. *HFP*—Hellenic Food Portal: The HFP will develop and maintain a database with available, registered and certified Greek products, which will be accessed via the Hellenic Food Quick Response Code (HFQRC) that will be developed for each product, or group of products and food-related services.

   d. *HFQRC*—Hellenic Food Quick Response Code: The code will be on food products and will connect to information on product and service labels at the portal HFP. The related information will be accessed by any consumer or professional, with a smartphone and internet access.

   e. *HFBF*—Hellenic Food Brand Franchise: A brand name of the Greek food of the highest quality must be established, developed and safeguarded as a franchise. This will be done in collaboration with government agencies of the GMAFRD. The brand name needs to be established as a franchise, so that free riding will be avoided, and only those firms with a serious commitment to the requirements of the franchise will participate and be granted the brand name.

f. *CDFM*—Centre for Development of Farmers' Markets: The status of farmers' markets needs to be upgraded and **professionalized**. The CDFM in collaboration with the NFFMC can facilitate this development.

g. *GFLH*—Greek Food Logistics Hub. This is a very key business activity that the GIPAFC must develop and facilitate. The GFLH will be developed initially in a key European commercial and trade centre in Europe, such as Rotterdam or Munich (the exact location must be chosen after careful research and a professional business plan). The GFLH will serve the export of Greek food products at large as well as at smaller scale. It will be able to link the demand and the European buyers, such as supermarket chains, brokers and other smaller buyers, with the suppliers of Greek food products in Greece. The GFLH must be designed and managed in a way to serve large, medium and even smaller scale exporters, SMEs and individual farmers, farmers' groups, cooperatives and federations of cooperatives.

h. *GAEUOF*—Greek Agriculture EU Office: The GIPAFC will establish and staff its own office in Brussels, which will provide intelligence and lobbying services on behalf of the Greek agri-food.

i. *Legal Office*: The GIPAFC will establish a legal office which will provide legal services and advice to its members, and individual farmers and other stakeholders for a fee. It will also provide services for conflict resolution between the various entities and agents of the agri-food chain. This is necessary, for larger firms, especially when international legal services might be necessary, as well as— and most importantly—for smaller agribusiness and farmers, who do not have the scale and resources for high-level legal representation and advice.

# 6.3   Some More Institutional Innovations

## 6.3.1   Food Ambassadors

After the financial crisis, the system of commercial attachés has deteriorated. Due to budget cuts, many embassies and consulates are understaffed, and often, the functions of promotion and intelligence of Greek exports are undertaken by non-specialized personnel. It is very important that the system of attachés is revamped and becomes more specialized and proactive. A network of well-trained "food ambassadors" should be established, with the role of collecting intelligence, networking and promoting Greek food exports.

- These skilled food ambassadors should either originate from the GIPAFC and be despatched with special assignments to key embassies and consulates. Or they should be highly skilled employees of the ministry but with some detachment and training at the GIPAFC and the GFLC.
- Food exports and tourism should be promoted in tandem at the Greek embassy levels.
- A system of assistant food ambassadors can be used for training students in the international agri-food markets.

## 6.3.2   Hellenic Agricultural and Food Research and Development Fund (HAFRDF)

The country and the agricultural sector need an independent research fund. The HAFRD must be created and operated under an independent board, under the auspices of the GIPAFC.

**Some of the branches of the fund:**

A. Basic biology and biotechnology research
B. Sectoral research, e.g. crops, trees, livestock and aquaculture.
C. Food innovation and development
D. Technology, engineering and information
E. Social sciences: economics and policy, sociology, management and marketing

## The sources of the fund:

A. A scaled national fee charged on the value-added of every agricultural product sold at the national market or exports. This fee will then be tax-deductible for the contributor.
B. A portion of the direct payments: This should not be a flat fee, but on a scale increasing on the share of the higher payments.
C. A flat portion of the fund from rural development
D. National funds from the secretariat of research
E. Research funds from the EU
F. Private donations
G. Other sources from national and international research competitions.

### Hellenic Agriculture and Food Scholarship Fund

The purpose of this fund will be to provide scholarships, to young scholars for graduate and postgraduate education, that will serve the development of Greek agriculture and food in the long run. An international committee will select among the best and most ambitious young scholars who will study abroad on topics and subjects related to agriculture, food and rural development. Both social and "hard" sciences will be supported. The scholars will be required to be involved in the Greek agri-food chain in some capacity and part time during their tenure.

## 6.4  Summary

A new paradigm for Greek agriculture will not materialize unless it is founded on a tangible, concrete organization which will ensure that the next steps are taken, that the value is created and that this value is distributed to all stakeholders in an efficient and equitable manner. This structure will have to build new and bypass existing inefficient, dysfunctional and anachronistic structures.

In this chapter we propose a concrete organizational and governance structure for Greek agri-food. Conceptually, the organization of the new paradigm for Greek agriculture is founded on a chain matrix that incorporates vertically, the agri-food chain; and horizontally four dimensions of organization: size, ownership, governance and geography. The new paradigm should incorporate small and large to account for diversity and for scale economies; private and public organizations where appropriate; atomistic and collective governance structures for functionality, efficiency and equity; and should be effective locally, regionally or internationally.

The core organization is the GIPAFC, the Greek Inter-Professional Agri-Food Council. An integrated, democratic and efficient structure. The GIPAFC is a square pyramid organization, with four pillars: (I) Inter-Professional and Inter-Branch organizations; (II) Cooperatives; (III) Research; (IV) Related industries.

The GIPAFC will not only be an industry representative or another collective lobbying agency but will also constitute a vital organization, which will coordinate, and further integrate vertically into organizations that will provide, research, extension, and other services to the industry.

# References

Damianos, D., and Vlachos. G. 2014. A National Strategy for Greek Agriculture (in Greek). In *International Crisis, Eurozone Crisis, and the International Monetary System*, ed. Chardouvelis and C. Gortsos, 179–190. Union of Greek Banks.

Kaditi, E.A., and Nitsi, E.I. 2010. Applying Regression Quantiles to Farm Efficiency Estimation. Report #112. EPE. Centre of Planning and Economic Research: Athens, Greece.

Matthews, A. 2012. Is selling experiences a potential growth path for European agriculture?. CAP reform. http://capreform.eu/is-selling-experiences-a-potential-growth-path-for-european-agriculture/. Accessed 2 Aug 2012.

Pine II, B.J., and J.H. Gilmore. 1999. *The Experience Economy: Work is Theatre and Every Business is a Stage*. Boston Massachusetts: Harvard Business School Press.

Schwab, M. 2013. The Global Competitiveness Report 2013–2014. World Economic Forum.

Swinnen, J., K. Van Herck, and T. Vandemoortele. 2012. The Experience Economy as the Future for European Agriculture and Food? *Bio-based and Applied Economics* 1 (1): 29–45.

# 7

# Epilogue

This book is a contribution to the quest for a new paradigm for Greek agriculture. I use the term "paradigm" in a loose sense as "an open-ended resource: a framework of concepts, results, and procedures within which subsequent work is structured" (Oxford Dictionary of Philosophy).

The book has five points of departure: Greek agri-food has shown remarkable resilience even in the dire conditions of the recent financial crisis (2008 to today); the value addition in Greek agri-food lags significantly behind its true potential and far from its competitors in other EU countries; there exist unexploited synergies with other sectors, such as tourism and logistics; the potential competitive advantage of Greek agri-food lies in its diversity which should not be treated as a constraint; the main impediments to growth and development of Greek agri-food are institutional in nature.

The need for a "paradigm shift" in the Greek agri-food industry follows the diagnosis that the problem is systemic rather than sectoral. It is the entire agri-food "system" that needs a revision, and not simply the industry, i.e. farm practices and structures, food processing and logistics. The constraints to the development of the agri-food industry can be found deep into the structure, history and idiosyncrasies of Greek

© The Editor(s) (if applicable) and The Author(s) 2017
K. Karantininis, *A New Paradigm for Greek Agriculture*,
DOI 10.1007/978-3-319-59075-2_7

agri-food, and the Greek polity. The new paradigm calls for a perspective that incorporates private and collective governance, large- and small-scale organizations, national and international scope, that is research-based and efficiency-driven.

The novelty of the "new paradigm" is a new organization. A by-pass. A scheme that goes around and beyond existing structures that have blocked the development and growth of the Greek agri-food. The proposal calls for a new concrete organizational scheme, the Greek-Inter Professional Agri-Food Council (GIPAFC). The GIPAFC will be a pyramid organization based on a comprehensive chain and network approach, that incorporates business organization, economic and political governance, economies of size, and inter-sectoral synergies. The GIPAFC will incorporate and nourish existing efficient structures and will by-pass those that are dysfunctional and anachronistic. The components of the agri-food chain are tightly linked and dependent on each other and require coordination and linkages to other industries, to research, and to local, national, European and international governance. Only an organization that encompasses all relevant stakeholders can play this role. The GIPAFC can be such an organization.

The book does not provide a direct and concrete policy recommendation. However, it provides a framework for the future national and European policy directions for Greek agriculture. The "new paradigm" provides the guidelines for the "carrot" and the "stick" for policy-makers: The future direction of subsidies and regulatory frameworks should have a vision for the future. The "new paradigm for Greek agriculture" provides a direction to where the incentives provided by subsidies should lead, and what limitations the regulations should put in place.

It is my deep conviction that Greek agri-food can become a driver and a model for the Greek economy if we all assist in unleashing its true potential. It is my wish, that this book will constitute a point of departure for a debate on the future direction and framework of Greek agri-food.

Uppsala, 2017

# Index

© The Editor(s) (if applicable) and The Author(s) 2017
K. Karantininis, *A New Paradigm for Greek Agriculture*,
DOI 10.1007/978-3-319-59075-2